CAMBRIDGE COUNTY GEOGRAPHIES

SCOTLAND

General Editor: W. Murison, M.A.

# ARGYLLSHIRE

AND

# BUTESHIRE

*Cambridge County Geographies*

# ARGYLLSHIRE
AND
# BUTESHIRE

by

PETER MACNAIR, F.R.S.E., F.G.S.
Curator of the Natural History Collections in the Glasgow Museums,
Formerly Examiner in Geology, University of Aberdeen

With Maps, Diagrams and Illustrations

Cambridge:
at the University Press
1914

CAMBRIDGE UNIVERSITY PRESS
Cambridge, New York, Melbourne, Madrid, Cape Town,
Singapore, São Paulo, Delhi, Tokyo, Mexico City

Cambridge University Press
The Edinburgh Building, Cambridge CB2 8RU, UK

Published in the United States of America by Cambridge University Press, New York

www.cambridge.org
Information on this title: www.cambridge.org/9781107657526



First published 1914
First paperback edition 2013

*A catalogue record for this publication is available from the British Library*

ISBN 978-1-107-65752-6 Paperback

# PREFACE

I AM indebted to Mr J. W. Reoch for his kindness in supplying the photographs on pp. 8, 23, 27, 31, 35, 52, 56, 57, 113, 130, 131, 134, 142, 144, 148, 155. For many valuable suggestions and aid in connection with the book I have to acknowledge the assistance of Mr David Gourlay, Mr James Park, Mr Alex Grey, and for the diagrams at the end that of my son, Mr P. Mackenzie Macnair.

P. M.

*April* 1914.

# CONTENTS

## ARGYLLSHIRE

## BUTESHIRE

# ILLUSTRATIONS

MAPS



The illustrations on pp. 11, 14, 51, 59, 63, 65, 68, 71, 78, 80, 83, 85, 88, 91, 98, 108, 110, 116, 139, 145, 150 are from photographs by Messrs J. Valentine and Sons, Dundee; that on p. 43 is from a photograph by Messrs F. Frith and Co., Reigate; the facsimile on p. 93 is taken from Dr Hume Brown's *History of Scotland*, Vol. II; and the portrait on p. 153 is reproduced from Mr J. A. Lovat-Fraser's *John Stuart, Earl of Bute* by kind permission of the author.

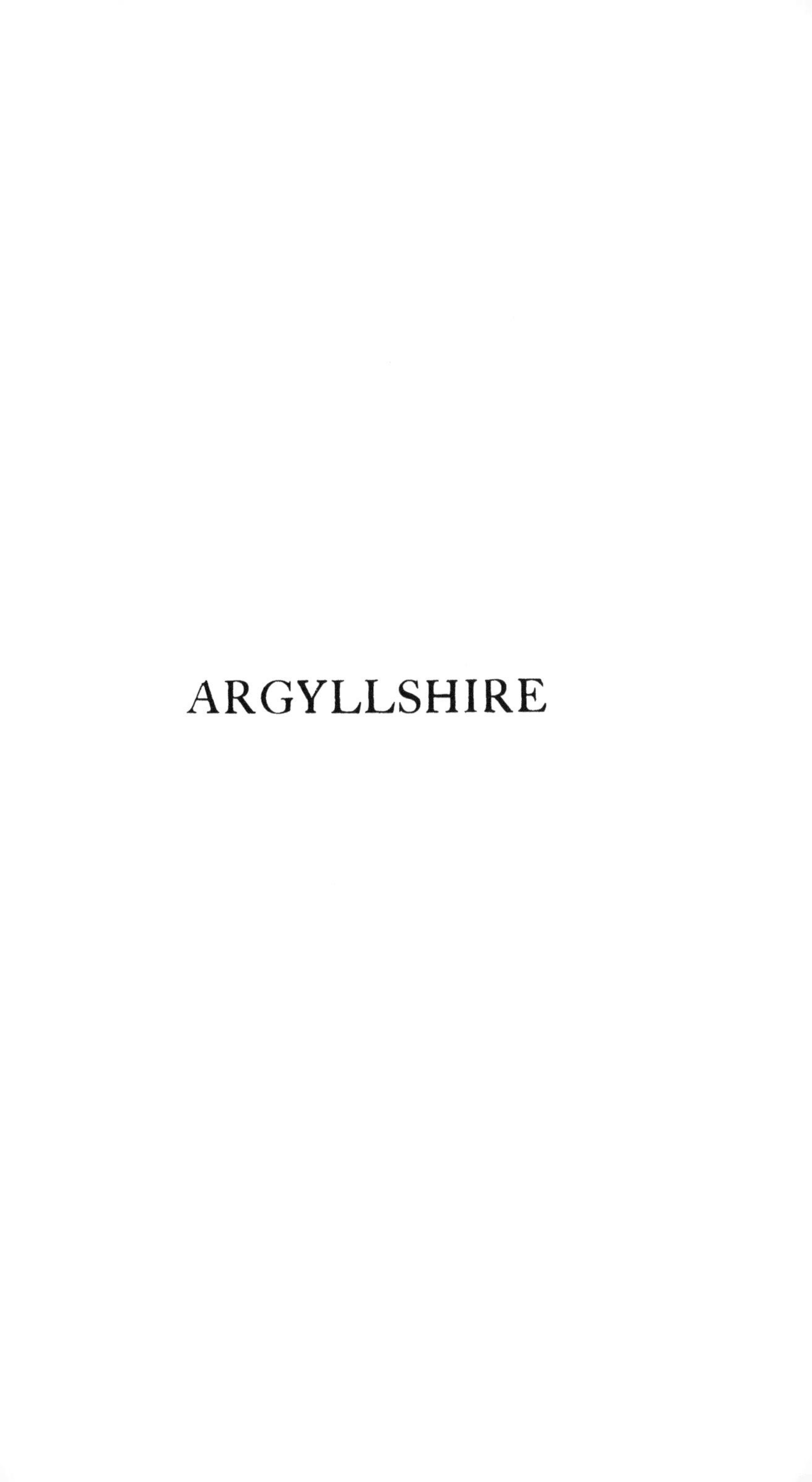

ARGYLLSHIRE

## 1. County and Shire. The Origin of Argyllshire.

The word *shire* is of Old English origin and meant office, charge, administration. The Norman Conquest introduced the word *county*—through French from the Latin *comitatus*, which in mediaeval documents designates the shire. *County* is the district ruled by a count, the king's *comes*, the equivalent of the older English term *earl.* This system of local administration entered Scotland as part of the Anglo-Norman influence that strongly affected our country after the year 1100. Our shires differ in origin, and arise from a combination of causes—geographical, political and ecclesiastical.

Argyllshire roughly corresponds to the old territory of the Dalriad Scots. Dalriada, however, at its largest, extended along the coast as far as Loch Broom; and the lordship of Argyll in the thirteenth century included the islands of the Clyde. It is frequently said that Argyll became a shire or sheriffdom soon after 1222, when it lost its independence, or at least after 1266. The Exchequer Rolls do not bear this out. There the earliest mention of a sheriff of Argyll seems to be under the year 1326.

In 1891 the Boundary Commission curtailed the county by adding to Inverness-shire the Small Isles and Kilmallie north of Locheil.

The county name comes from the name of one district, and is regarded as being *airer gaedhil*, "the land of the Gael."

## 2. General Characteristics and Natural Conditions.

Argyllshire is a maritime county, no part being more than 12 miles distant from either the open sea or some sea loch. It is at once the most southerly and the most westerly of the Highland counties. This with the deeply indented character of its coast line places much of the county in a somewhat disadvantageous relation with the rest of the country as regards connection by land; but that is to some extent counterbalanced by the fine steamer service which the coastal parts now enjoy.

The coasts and sea lochs present an abundance of wild and picturesque scenery. As we sail down the Firth of Clyde, many beautiful vistas open into the very heart of Cowal by sea lochs like Loch Long, Loch Goil, the Holy Loch and Loch Riddon, which are screened on either side by lofty rugged mountains. We enter the Kyles of Bute by what appears to be a capacious bay, and then gradually the land closes in on each side till the sea-way looks more like an inland river flanked by steep hills and rocky banks. In the lower parts of Loch Fyne the scenery is somewhat tame, but the higher parts are encircled with lofty mountains. Sailing up the Firth of Lorne and Loch Linnhe and round the islands, we find displayed a magnificence of scenery unsurpassed by anything in the kingdom. Everywhere the side which borders the sea is traversed by deep bays and creeks winding in a variety of directions so as to form the land into a series of peninsulas and islands.

The interior of the county is equally diversified. Its general appearance is wild and mountainous, especially in the northern parts bordering on Perthshire and Inverness-shire. Many of the loftiest and most massive heights of the Scottish Highlands and many of its longest and deepest glens are to be found in Argyllshire, while large areas are covered with tabular moorlands. Here and there we find fertile valleys, as Glendaruel, with sometimes a considerable amount of arable land. Many of the valleys, as Glencroe and Glencoe, are really alpine in their character and impressively sublime. Glencoe, says Macaulay, "is the most dreary and melancholy of all the Scottish passes, the very Valley of the Shadow of Death. Mists and storms brood over it through the greater part of the finest summer; and even on those rare days when the sun is bright, and when there is no cloud in the sky, the impression made by the landscape is sad and awful. The path lies along a stream which issues from the most sullen and gloomy of mountain pools. Huge precipices of naked stone frown on both sides. Even in July the streaks of snow may often be discerned in the rifts near the summits. All down the sides of the

**Glencoe**

(*From a picture by Horatio Macculloch, R.S.A., Corporation Art Gallery, Glasgow*)

crags heaps of ruin mark the headlong paths of the torrents. Mile after mile the traveller looks in vain for the smoke of one hut, or for one human form wrapped in a plaid, and listens in vain for the bark of a shepherd's dog, or the bleat of a lamb. Mile after mile the only sound that indicates life is the faint cry of a bird of prey from some storm-beaten pinnacle of rock."

Scotland is divided into three parts—the Highlands, the Southern Uplands, and the Midland Valley. The dividing line between the Highlands and the Midland Valley, known as the great Highland boundary fault, is a fracture in the earth's crust, which crosses Scotland from shore to shore with a north-east and south-west trend. It can be traced through the Mull of Kintyre, from which it crosses to the west side of Arran near Dougrie and curves round to the extreme north end of the island; then crossing the Sound of Bute, it appears on the Island of Bute at Scalpsie Bay and trends across that island by way of Loch Fad to Rothesay Bay. Passing under the sea, it reappears on the mainland at Toward Point and stretches across the Cowal peninsula to Inellan Pier. Again passing into the sea it crosses to Kilcreggan and enters Dumbartonshire. From this point it has been traced across the whole of Scotland to Stonehaven.

Practically the whole of Argyllshire lies to the north-west of this great line and is therefore both geologically and geographically a portion of the Scottish Highlands, while Arran and the Island of Bute are both divided into two portions by this line.

## 3. Size. Shape. Boundaries.

The county of Argyll lies between 55° 15′, and 56° 55′ north latitude and 4° 32′ and 6° 56′ west longitude. The mainland portion approaches the shape of a triangle, of which a line running from the Point of Ardnamurchan along the borders of Inverness-shire to the moor of Rannoch forms the base and another line running from thence to the head of Loch Long and along the Firth of Clyde one of the sides, the third side being the shores of the Atlantic ocean. Besides this, Argyllshire includes a large number of islands, the chief being Mull, Islay, Jura, Tiree, Coll, Rum, Lismore and Colonsay. The county is estimated to have a circumference of 460 miles and its area is 1,990,471 acres or 3110 square miles. Its length is little short of half the length of the west coast of Scotland.

Argyll is bounded on the north by Inverness-shire, on the east by Perthshire and Dumbartonshire and by the ramifications and main expanse of the Firth of Clyde, on the south by the Irish Sea, on the west by the Atlantic ocean. Beginning at Ardnamurchan Point, the boundary line strikes eastwards by the south shores of Loch Moidart and Loch Shiel to the head of the latter loch. From here it goes east along Loch Eil and then south till opposite the entrance to Loch Leven, when it strikes due east by the south shore of Loch Leven to the head of the river Leven at the meeting point of the counties of Argyll, Inverness and Perth. From the headwaters of the Leven the boundary line marching with Perthshire passes in an irregular line almost due south, curving round the head of Loch Laidon and passing over the summits of Carn Daimh (2291 feet), Ben Achallader (3399), Ben a Chaisteil (2897), and Ben Odhar (2948), till it reaches the watershed between the river Lochy and the river Fillan. It then mounts to the summit of Ben Laoigh (Ben Luï) and Meall nan Caord (2568 feet). Still striking southward, it passes over the high ground to the west of Loch Lomond, crossing Ben Vane at an altitude of 3004 feet, and descending Glen Loin reaches the sea at the head of Loch Long. From thence it circles round the Cowal peninsulas to the head of Loch Fyne, and then descending the west side of that loch it passes round the peninsulas of Kintyre and Knapdale to the mouth of the Crinan Canal at Crinan. The boundary now proceeds due north in an irregular line to Loch Leven and circles round the peninsulas of Morven and Ardnamurchan to the point where we began. The islands which belong to the county are everywhere bounded by the waters of the Atlantic ocean.

## 4. Surface and General Features.

The region embraced by the mainland of Argyll may here be defined as the south-western prolongation of the Grampian Mountains, the sea end of which is represented by a vast archipelago. In some cases where submergence has not been sufficient to produce complete isolation the sea has broken down the weaker barriers and penetrated far inland along the lines of pre-existing valleys.

The most mountainous part of the country is situated in the north-east, where the Grampians have reached their greatest height as a range and fall gradually into the Atlantic. Some of the more lofty and more conspicuous

The Cobbler

summits are as follows: Bidean nan Bian (3766 feet), between Glencoe and Glen Etive, Ben Laoigh (3708 feet), on the boundary line between Argyll and Perth, Ben Cruachan (3689 feet), between Loch Etive and the foot of Loch Awe, Ben Starav (3541), east of the head of Loch Etive, Ben a Bheithir (3392), to the south-west of Ballichulish, Buchaille-Etive (3345), near the top of

Glen Etive, Culvain (3224), on the northern border, Cobbler (2891), Loch Long. The altitude declines towards the sea into such heights as Bishop's Seat (1651 feet) above Dunoon, Cruach-Lussa (1530) eastward of Loch Sween, and Ben-an-Tuirc (1491) in Kintyre. In the islands there are several important mountains, as Benmore (3185 feet), in Mull, and the Paps of Jura (2365).

Viewed from one of the higher altitudes, the Highlands of Argyll are seen to form a great level plateau deeply furrowed with valley systems, the whole area being but a much dissected portion of the main Highland plateau. A few such *massifs*, such as Ben Nevis, raise themselves above the general level, and these may represent ancient monadnocks or islands which were never reduced to the general level of the plateau. This tableland is not due to the elevation of horizontal beds, but represents a base-level of erosion which has been carved into its present appearance by the action of the denuding agents.

## 5. Watershed. Rivers. Lakes.

The main watershed of this district of the west Highlands lies in some parts within and in some outside the county boundary, while for a considerable distance, where Argyllshire marches with Perthshire, the watershed coincides with the boundary line.

Starting in Inverness-shire between Loch Nevis and Loch Morar, the watershed runs generally eastward to near the falls of the river Pattock, its most easterly point. Then it strikes south-west, skirting Ben Alder, and crossing the summit of Carn Dearg, reaches the high road at the head of Glencoe near King's-House Inn. Taking a sudden bend, it mounts the high ground which surrounds the headwaters of Glen Lyon, and then passing through Ben Achallader and Ben Odhar it falls to Tyndrum, Perthshire, where it separates the Lochy and the Fillan.

From this point it rises to the summit of Ben Laoigh and strikes along the high ground that looks down upon the waters of Loch Awe to the west and the waters of Loch Fyne to the east. It then divides the districts of Knapdale and Kintyre from end to end, passing through the Crinan Canal and between East and West Loch Tarbert, finally reaching the sea at the Mull of Kintyre.

The rivers and streams of Argyllshire are all short and rapid, many of them being merely torrents descending steep and narrow glens. The principal of these follow the south-west trend of the country, which coincides with the strike of the rocks entering the great sea lochs at their heads, while innumerable mountain torrents descend the mountain ridges, reaching the sea lochs by the nearest route. The former type lie in what has been termed the longitudinal valleys, the latter in the transverse valleys.

Taking the longitudinal valleys in order from south to north, we have first Loch Long, which receives at its head the small mountain rivulet, the Loin Water.

A similar stream rises near Ben Laoigh, and after a course of about $6\frac{1}{2}$ miles reaches the sea near Cairndow at the head of Loch Fyne. Between the head of Loch Fyne and the foot of Loch Awe lie the Shira and Aray, which flow southwards into Loch Fyne near Inveraray. The Shira rises on Ben Bhuidhe at an altitude of 2760 feet and has

Loch Riddon

a course of eleven miles. The Aray, which rises near the watershed between the head of Loch Fyne and the foot of Loch Awe, has a course of eleven miles. The stream runs at first on a rocky bed and between bare hills, but as it descends its sides become richly wooded. It passes over several picturesque waterfalls, perhaps the most notable being that about three miles above Inveraray, where the stream leaps down a precipice 60 feet in height. The next longitudinal valley is that of Loch Awe, which receives at its foot or north-east end the Strae and the Orchy, the latter the longest stream in Argyllshire. The Strae rises at an altitude of 1250 feet, flows south-westward for 8½ miles and joins the Orchy about half a mile above its influx into Loch Awe. The river Orchy rises on the main watershed close to the Perthshire boundary at an altitude of 2700 feet and flows south-westward for 10¾ miles under the name of the Tulla Water, ultimately expanding into Loch Tulla. Here it receives an important tributary from the mountains which flank the south-west side of the loch. After emerging from the loch it runs for 16½ miles in a south-westerly direction, joining Loch Awe at Kilchurn Castle. At first the stream has a turbulent course along the valley of Glenorchy, but upon reaching the Vale of Dalmally it becomes more sluggish. Still further north is the valley of Loch Etive, which receives at its head the waters of the river Etive. This stream issues from Lochan Mathair Etive close to the watershed and has a course of 15¼ miles flanked on the right by Buchaille-Etive and Bidean nan Bian, and on the left by Clach Leathad and Ben Starav. The next important valley is that of Loch Leven, into the western end of which runs the river Leven. This stream rises in a small loch close to the watershed and flows westward 16¾ miles through a chain of lochs amid wild and romantic scenery. In the districts of Ardgour and Morven there are several heavy mountain torrents, and the only streams of any importance in the islands are the Aros Water in Mull and the Laggan in Islay.

The freshwater lochs are exceedingly numerous, and vary in length from upwards of twenty miles down to small ponds. Loch Awe, the longest of all, runs in a south-west and north-east direction from Ford in Nether Lorne to the foot of Ben Cruachan and Dalmally. In extent and picturesqueness it is one of the most important of the freshwater lakes of Scotland. The outflow by the deep gorge of the Pass of Brander is held to be of comparatively recent origin. Anyone, according to Geikie, going from Loch Crinan to Kilmartin and then up the terraced valley to the south end of Loch Awe, must be convinced that this is the old outlet. At its north-eastern end round the foot of Ben Cruachan it expands into two great branches and is studded with many beautiful islands. On one of these stand the ruins of Kilchurn Castle built in 1440 by the Breadalbane of that day, who was styled the Black Knight of Rhodes. Other freshwater lochs are Loch Avich, Loch Laidon, Loch Eck, Loch Arienas, Loch Nell, Loch Avisa, Loch Shiel, and Loch Tulla. The sea lochs will be described in the section dealing with the coast line. In the meantime it may be stated that such fiords as Loch Fyne, Loch Long and Loch Goil are merely submerged valleys, and many of them have been demonstrated by the Admiralty soundings to be true rock basins which have been excavated by glacier ice. If the sea board of Argyllshire were to be sufficiently elevated, it would be found that the region now covered by the sea would present an analogous appearance to that on

The Pass of Brander, Loch Awe

the eastern slopes of the mainland, namely, deep mountain-glens with rock-bound lakes.

Classified according to their origin and mode of occurrence the lochs of Argyllshire can be arranged in three distinct types. The first have been hollowed out of the solid rock and are known as true rock basins. These include all the larger and more important sea and freshwater lochs such as Loch Fyne and Loch Awe. It is now generally admitted that these rock basins, either as arms of the sea or freshwater lochs, have been formed by differential ice erosion acting upon highly folded beds of unequal hardness along the geological grain of the country; more rarely, however, they lie across the grain. Second are those which have been formed by the ponding back of a sheet of water by glacier debris. This type is usually confined to the heads of glens or the mouths of corries. The third type includes all those which lie in cup-like hollows either in the old glacier moraines or in the boulder-clay.

## 6. Geology and Soil.

Geology is that department of science which deals with the past history of the earth as it is recorded in the rocks. It seeks to show how each continent and island has been built up, how seas, lakes, and rivers have been formed and how the various useful products found in the crust of the earth have been accumulated. Its great ideal is to restore in imagination the geographical and physiographical features of each successive period in geological time and to reclothe them with their long-vanished plants and animals. The great rock masses of which the earth's crust is composed can be classified into two kinds according to their mode of origin—the igneous or those that have been erupted from the interior of the earth in a molten condition, and the aqueous or those that have been formed as sediments at the bottom of seas and lakes. The latter are also called sedimentary rocks.

The igneous rocks have been divided into two groups—plutonic and volcanic. The plutonic rocks have cooled at some distance below the surface and have solidified much more slowly than the volcanic. Hence they have assumed a more coarsely crystalline structure. They commonly occur in great intrusive bosses. The plutonic masses are represented by the granite of Ben Cruachan and the Ross of Mull and by numerous intrusions of diorite. The volcanic rocks comprise those which have been ejected upon the surface of the earth by volcanic action, and have been laid down either as great sheets of lava, or accumulations of lava and volcanic dust. Lavas of Old Red Sandstone Age occur near Oban and the Pass of Brander, while lavas of Tertiary Age occur in Staffa, parts of Morven, Mull and Ardnamurchan.

The aqueous or sedimentary rocks include all those which, like sandstone, have had a secondary or derivative origin or, in other words, have been formed out of previously existing materials, as well as a few others which, strictly speaking, do not answer to this description of their origin. The fragmental rocks are those which owe their origin to moving water, such as gravel, sand, or mud which become consolidated into conglomerate, sandstone and shale. Another great division of the sedimentary rocks comprises the organically formed rocks, which have been built up by the slow accumulation of the remains of plants and animals. Coal and limestone are familiar examples of this class. Sedimentary rocks of Upper Old Red Sandstone and Carboniferous Age occur at the south end of the peninsula of Kintyre and of Jurassic Age on the south shores of the promontory of Ardnamurchan.

Geologists have laid down the order of succession of the stratified formations as follows:

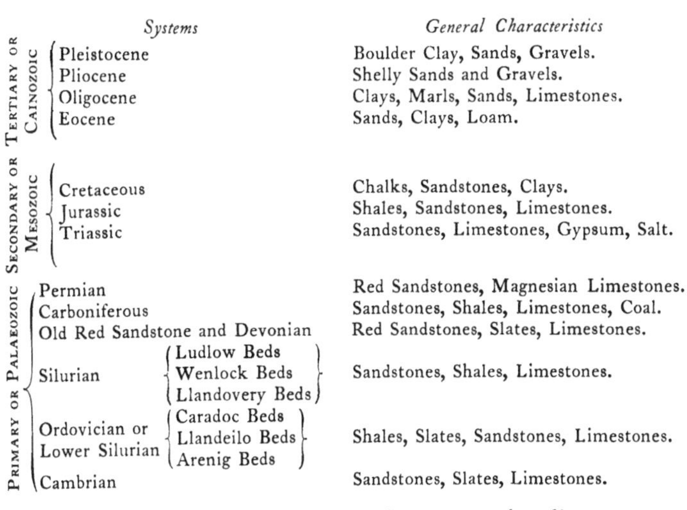

| | *Systems* | | *General Characteristics* |
|---|---|---|---|
| Tertiary or Cainozoic | Pleistocene | | Boulder Clay, Sands, Gravels. |
| | Pliocene | | Shelly Sands and Gravels. |
| | Oligocene | | Clays, Marls, Sands, Limestones. |
| | Eocene | | Sands, Clays, Loam. |
| Secondary or Mesozoic | Cretaceous | | Chalks, Sandstones, Clays. |
| | Jurassic | | Shales, Sandstones, Limestones. |
| | Triassic | | Sandstones, Limestones, Gypsum, Salt. |
| Primary or Palaeozoic | Permian | | Red Sandstones, Magnesian Limestones. |
| | Carboniferous | | Sandstones, Shales, Limestones, Coal. |
| | Old Red Sandstone and Devonian | | Red Sandstones, Slates, Limestones. |
| | Silurian | Ludlow Beds<br>Wenlock Beds<br>Llandovery Beds | Sandstones, Shales, Limestones. |
| | Ordovician or Lower Silurian | Caradoc Beds<br>Llandeilo Beds<br>Arenig Beds | Shales, Slates, Sandstones, Limestones. |
| | Cambrian | | Sandstones, Slates, Limestones. |

It is often found that both igneous and sedimentary rocks have been altered by pressure or by coming into contact with molten igneous material. In this way clay-slate or shale may be altered into slate, and sandstone into quartzite, while a shaly sandstone may pass into a mica schist. Such igneous rocks as granite become gneiss and whinstone is altered into hornblende-schist. When rocks have been subjected to such alterations, they are known as metamorphic rocks.

The mainland of Argyll and several of the more important islands such as Islay and Jura consist chiefly of the metamorphic rocks of the eastern Highlands. In this region sedimentary rocks have been altered by metamorphism from such normal sediments as conglomerates, sandstones, shales and limestones into schistose conglomerates, quartzites, slates and crystalline limestones. These sediments prior to their metamorphism were penetrated by intrusive igneous rocks, which have also suffered in the general metamorphism passing into gneisses and hornblende schists. At a later period the metamorphic rocks were invaded by great masses of igneous material principally granite, which produced a still further stage of metamorphism along the line of contact.

The sedimentary rocks of which the Highlands are composed were no doubt originally formed in more or less horizontal beds. They are no longer seen to occupy their original horizontal position but have usually been bent into a series of folds due to a great succession of earth movements which culminated during the formation of the rocks of the Lowland Plain. These movements seem to have been due to pressure applied from the north-west and south-east, as a result of which the solid rocks were buckled into a complex series of long arches and

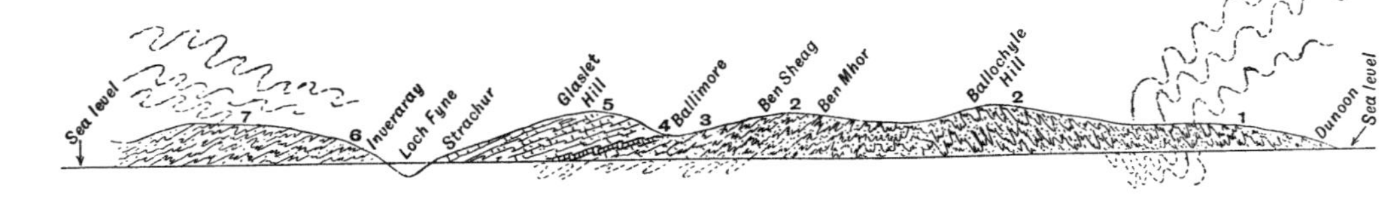


**Section from Glen Aray to Dunoon**

1. Dunoon Phyllites. 2. Ben Ledi Grits. 3. Green Beds. 4. Loch Tay Limestone. 5. Garnetiferous Schists. 6. Ardrishaig Phyllites. 7. Central Highland Quartzites

troughs like the waves of the sea. These earth waves naturally arranged themselves at right angles to the lines of pressure, and so we find a marked tendency in the valleys and corresponding mountain ridges to take a north-east and south-west strike.

The different schists or metamorphic rocks which form the Highlands of Argyllshire traverse the county in bands or zones having a general north-east and south-west trend and may be said to be roughly parallel with the great boundary fault. A brief description of these various schist zones, proceeding in order from the shores of the Firth of Clyde in a north-westward direction, will now be given.

Along the shores at Dunoon occurs a group of slates and phyllites which are referred to the lowest position in the whole series. These are followed towards the north-west by the Ben Ledi grits, which form by far the greater part of the highest ground of Cowal lying between the Firth of Clyde and Loch Fyne, as well as a large part of Kintyre. In Glendaruel and along the shores of Loch Fyne the latter group is succeeded by the Green Beds, the Loch Tay Limestone and the Garnetiferous Schists, which are prolonged into Kintyre.

The next important belt is the Ardrishaig phyllites, varying in width from 3 to 6 miles. It has been traced from Glen Shira along both shores of Loch Fyne to Loch Killisport in North Knapdale. Then follow the Easdale Slates with the dark limestones, the main Limestone of Loch Awe and the Schiehallion Quartzites, which are repeated by innumerable folds spreading northwards along the shores of Loch Linnhe and stretching westwards to Jura and Islay. An interesting feature of this series is the "Boulder Bed," a conglomerate lying near the base of the quartzites and made up of rounded fragments of the underlying rocks and of granite, the latter not now found in that district. The Ardrishaig phyllite series, which appears along the shores of Loch Fyne, passes below the Loch Awe Quartzites in a complex trough and reappears at Craignish near Kilmartin.

Associated with these metamorphic rocks of sedimentary origin are bands of epidiorite, of igneous origin. Being intruded into the sedimentary rocks prior to the general metamorphism of the region, they have suffered alteration along with the sedimentary rocks. Sills of epidiorite are numerous in the Loch Awe, Knapdale and Islay districts, and rocks of Torridonian age occur in the island of Islay.

Rocks belonging to the Old Red Sandstone system cover a large area in Lorne between Loch Melfort, Oban, and the Pass of Brander. They also appear among the lofty mountains of the Glencoe district. The rocks consist for the most part of lavas and tuffs, resting uncomformably upon the older metamorphic rocks. The plutonic equivalents of these lavas are represented by the granite of Ben Cruachan and the diorite of Glen Domhain. The volcanic plateau of Lorne is intersected by numerous dykes of porphyrite, which also cut through the schists and granite of Ben Cruachan.

Rocks belonging to the Upper Old Red Sandstone formation occur in isolated patches along the shores of the Firth of Clyde, on the west coast of Kintyre, and between Campbeltown and Southend. They rest un conformably upon all rocks older than themselves, and at Campbeltown they pass upwards into the volcanic rocks of Lower Carboniferous Age, which are succeeded by the lower limestones and coals of the Lower Carboniferous Limestones.

Small patches of Jurassic strata occur on the north and south shores of the promontory of Ardnamurchan. These range from the Lower Lias to the Oxford Clay, while representatives of the Upper Greensand covered by basalts of Tertiary Age are found in Morven and on the shores of Loch Aline.

The great basaltic plateaus of Antrim and the Inner Hebrides, whose lava beds are so extensively developed in the Island of Mull and which make such a picturesque feature in the far-famed Staffa, are considered to belong to the Oligocene division of the Tertiary Period. The evidence for the age of these lavas is principally derived from the leaf beds which occur at Antrim and at Ardtun[1] in Mull. The plants appear to have grown in pond-like hollows on the surface of the lava flows during a period of volcanic quiescence and were buried beneath the succeeding outpourings of igneous material. The leaves are those of plants allied to the yew, rhamnus, plane and alder.

According to Professor Judd we have in the islands of Skye, Mull, Rum and St Kilda the denuded stumps of great volcanoes, that of Mull being on a grander scale than the modern volcanoes of Italy. Associated with

[1] Discovered by the late Duke of Argyll.

this phase of vulcanism there are innumerable basalt dykes which have been traced from Yorkshire to Orkney and from the north of Ireland to the east coast of Scotland.

Abundant evidence occurs throughout the county of the long period of glaciation which has stamped with more or less distinctness its influence upon the physical

**Moraines in Strath of Orchy, and Ben Cruachan**

features. Many of the Highland valleys are beautifully smoothed and polished in the direction taken by the ice. Travelled boulders are found all over the region and fine examples can be seen of the moraines formed during the later valley glaciation. After the glaciers had shrunk back to the valleys, the sea stood at a level of about

100 feet higher than at present, when the stratified clays and sands containing arctic shells were slowly deposited.

The great variety of the geological formations accounts for the variety of soils in Argyllshire. Light gravelly soil and sandy loam are the most prevalent. Peat mosses are also widely distributed, while here and there we find a few meadows composed of rich alluvial soil. It has already been pointed out that owing to the configuration of the land the stream courses are but short and as a consequence no extensive or deep haughlands occur. The nature of the soil is of course largely dependent upon the character of the underlying rock. There are, however, exceptions to this rule. Thus, the granite in the Glen Etive district makes a very poor soil, while that at the head of Loch Sunart yields remunerative crops. The lavas of the Morven district yield many rich grasses, while those of Glen Etive and Glen Kinglass are almost barren. As usual, wherever limestone occurs, the soil is rich and productive. A fine example of this is seen in the island of Lismore. The decomposition of clay slate also makes an excellent soil, as in the islands of Seil and Luing.

## 7. Geology and Scenery.

An intimate relationship exists between the geological structure and the scenery of the county. In considering this, it must be kept in mind that the main features of the Argyllshire scenery have been produced by the dissection of the Highland plateau by the ordinary agents of subaerial denudation.

The valley systems of the Highlands may be classified into two main types—the longitudinal and the transverse; the former lying parallel with the general trend of the mountain chain, the latter crossing the longitudinal valleys approximately at right angles. Examples of the longitudinal valleys are Loch Long, Loch Fyne, Loch Awe and Loch Linnhe; and of the transverse valleys Loch Goil, Loch Eck, Loch Striven, the Kyles of Bute, and most of the minor cross streams. The transverse valleys would have their initial direction given to them by the slope of the marine plain of denudation towards the south-east, while the longitudinal valleys would be determined by the geological structure of the ground.

The present drainage system of this ancient high plateau was most likely initiated in Tertiary times and subsequent to the outpouring of the basaltic lavas of the Hebrides. This seems to be clearly demonstrated by the manner in which such valleys as Loch Long, Glen Tarbert on the west side of Loch Linnhe and others cut through Tertiary basalt dykes.

Many of the Argyllshire valleys of the transverse type are but relics of a once continuous drainage system leading more or less from west to east. This system, however, has to a great extent been broken up and segmented by subsequent valleys of the longitudinal type, which have been developed along lines of special weakness. In the Cowal district, Loch Goil may be taken as a good illustration of a transverse valley cut through by the longitudinal valley of Loch Long, its south-eastern prolongation being the present valley of the Gareloch. In the north of the shire the Lairigmor valley and river Leven valley seem clearly to have been continuous though now separated by the valley of Loch Leven. After a river system has begun to be broken up along lines of structural weakness and secondary watersheds produced, it may quickly assume a very complex drainage system.

As we sail down the Firth of Clyde we cannot help observing the smooth outline of the hills in the immediate foreground, such as rise above Dunoon, Blairmore and Inellan. But if we look up the Holy Loch or Loch Long towards the centre of Cowal we are at once struck with the rough and jagged outline of the mountains as seen against the skyline. The rougher area corresponds with one of the great axial lines of folding into which the rocks have been thrown, and there can be no doubt that the highly altered and crumpled character of the schists along this line has to a large extent determined the rugged character of the scenery It ought of course to be borne in mind that the higher summits are free from drift, and as it is only these portions of the mountains that can be seen from the distance this will also help to accentuate their wild appearance. On the other hand, the hills in the foreground can be seen from top to bottom and the eye is at once arrested by their drift-covered slopes.

In the north-east portion of the county perhaps the most conspicuous feature in the landscape is the granite mass of Ben Cruachan, which forms a well-marked east and west ridge and rises into two sharp peaks. To the north-east of Ben Cruachan and in this same granite track, a series of deep glens surrounded by wild and lofty mountains forms a practically inaccessible country. On the west side of Loch Etive a similar group of granite hills rises to an elevation of over 3000 feet in Beinn Sguliaird. Everywhere throughout this district the scenery is typically that of a granite region.

In the volcanic district of Lorne the scenic features

Corrie on Ben Cruachan

are entirely dissimilar. The forms of the hills and lines of crag and escarpment are determined by the inclination, succession and general character of the lava-flows and masses of fragmental rock of which they are composed. This type of scenery is well shown in the neighbourhood of Oban.

The effects of glaciation in rounding and polishing

the rocks is everywhere observable, while the large mounds of moraine material often lend a rugged aspect to the foreground. Within the last few years it has been claimed that a valley glacier has the power of forming a cirque or corrie near the head of the valley. The great factor in the production of the cirque is the large gaping crevasse, usually of great depth, which is formed where the body of the glacier moves away from the snowfield, the rapid recession of the cirque being the result of the sapping through daily summer frost work of the rock at the base of the crevasse. This view has recently been applied to explain the origin of the corries in the Scottish Highlands.

## 8. Natural History.

Scotland is included in that great zoological province known as the Palaearctic, which extends from Ireland to Japan and comprehends within its boundaries the temperate regions of the eastern hemisphere. Over the whole of this vast area the fauna shows a generic similarity which is sufficiently distinct to mark it off from the other zoological regions of the globe. The different parts of the province exhibit, however, much specific diversity between its plants and animals. Argyllshire contains but few plants or animals that are not to be found in other parts of Great Britain. But it does not follow that they are to be found in all parts of the island. Thus some species have their northern limit while others have their southern, eastern, or western limit within the county.

It is now generally believed that the greater part of the British fauna and flora reached these islands by a land connection with the Continent. From evidence which cannot be discussed here it is supposed that towards the close of the Ice Age the British Isles underwent a slow upheaval to a height probably corresponding to the 80-fathom line, the consequence being that the present bed of the North Sea was elevated into land, through which flowed the Rhine with the Thames, Ouse, Tay and other British rivers now entering the North Sea as its tributaries. At this time the English Channel and the Irish Sea formed a group of low-lying grounds uniting Britain and Ireland to the Continent so that the immigration of the arctic-alpine flora and fauna took place step by step across the plains from these centres of dispersion till they covered the whole of the British Isles.

Towards the close of glacial times, when the great ice sheet had passed away and only local glaciers were to be found here and there in the mountainous districts, the low grounds of Central Europe were covered by an arctic-alpine flora and fauna. With the gradual amelioration of the climate these plants and animals were forced to retreat to higher latitudes, while those inhabiting Central Europe retreated to the higher mountains, closely followed by the incoming march of the temperate species. There can scarcely be any doubt that it was the arctic-alpine flora that first covered these islands after the retreat of the glaciers. The commonest animals in Britain at that time were the reindeer, the elk, the mammoth, the wolf, and so forth. After the retiral of these northern plants and animals to higher latitudes, the country was invaded by a temperate flora which is now the prevalent type of vegetation.

It is impossible to say how long the land remained at this high level, but there is strong evidence to show that when the existing fauna and flora migrated into Britain the country was undergoing a gradual subsidence. As a result of this Ireland was first of all separated from England and at a later period England was separated from the Continent. The earlier separation of Ireland from Britain explains the comparative paucity of mammals and reptiles in the former country. That is, Ireland had been cut off before these animals began to migrate into England.

For botanical purposes this region may be divided into Alpine, sub-Alpine, lowland and littoral districts. In the Alpine district such species as *Salix herbacea*, *Vaccinium myrtillus*, *Empetrum nigrum*, *Alchemilla alpina* are of common occurrence. The rarer Alpine plants in Perthshire are confined to a certain band of schist, known as the Ben Lawers phyllites. This band enters Argyll on the eastern border at Ben Laoigh, where many of the rarer forms are to be found. The band of schist then descends rapidly to the level of the sea and soon reaches altitudes unsuitable for the existence of an Alpine flora. In the sub-Alpine zone *Calluna* and its grass associations extend over most of the area. The tree associations of this region are pine, birch and oak, the bracken being characteristic of its lower border. In the lowland district the familiar plants are found, including many of the cultivated plants introduced by man. Whin and broom are also characteristic of this region as well as many of the broad-leaved perennial herbs and woods of mixed trees including the beech. The littoral district is inhabited by the salt-loving plants and by the marine associations. Many of these belong to a peculiar Atlantic type, such as *Carum verticillatum*, *Brassica monensis*, *Cotyledon umbilicus*, *Asplenium marinum*.

Ben Laoigh, from Dalmally

We pass now to consider the fauna of the region. Several species of bats have been recorded. Among the Insectivora the hedgehog, the mole and the common shrew are abundant. The land Carnivora are represented by the wild cat, which is totally absent from the isles, and although not extinct on the mainland, has receded to the least accessible districts. The weasel and the stoat are plentiful, but the marten and the foumart are local and becoming very scarce. The otter is abundant on the mainland and the isles. The badger lives on the former only, but less so than in the past. Of Scottish mammals the marine Carnivora, represented by the seals, are undoubtedly the most interesting to the naturalist. We have as constant residents in the Argyllshire waters the common seal (*Phoca vitulina*) and the larger and more local grey seal (*Halichaerus grypus*), which is distinctly oceanic and insular, frequenting for the most part the more remote Hebrides. Others which may or may not be of casual appearance are the ringed seal (*Phoca hispida*) and the Greenland or harp seal (*Phoca groenlandica*). The Cetacea include the common rorqual, the lesser rorqual and the white-sided dolphin. The red deer and the grouse are perhaps the most important members of the Scottish fauna, especially when we look at them from the standpoint of the amount of money which they bring into the country. The fallow deer exist only in a semi-domesticated state in parks or on some of the islands. Roe deer are common in Islay and Mull and in some of the most suitable parts of the mainland, especially in the comparatively lately planted areas of any size, such as the larch and fir woods of Ardnamurchan and Strontian. Among the rodents the following may be noted as natives of the shire: the squirrel, the brown rat, the common mouse, the long-tailed field mouse, the common field vole, the common hare, the mountain hare and the rabbit.

Of the 368 species of the British avi-fauna, 210 have been ascertained to be resident or occasional visitors to Argyll and the Isles. The recent extension of woods and plantations has favoured the increase of some of the species, as the resident *Turdidae*. The redstart has increased largely in recent years as well as the spotted fly-catcher, wood wren, grasshopper warbler, starling and jackdaw. The aggressive jackdaw is answerable for the diminution of the old-fashioned chough. The rook is increasing and seems to be developing both its carnivorous and its destructive propensities. The raven also appears to hold its own against other competitors. On the other hand, the goldfinch is becoming both local and rare, and the yellow-hammer is said to be rapidly decreasing, especially in the neighbourhood of Inveraray. The lesser white-throat, chiff-chaff and yellow wagtail are said to be of doubtful occurrence. It would be impossible to refer to the birds of the county in detail, and we may simply state that the game birds, owls, eagles, falcons, hawks, geese, ducks, waders, gulls, terns and divers are fully represented by the ordinary British species. Of the Reptilia, the adder is general in Jura, Mull and the western part of the Ardnamurchan peninsula; the common lizard is widely dispersed over the moorland localities of the mainland; and the slow worm is distributed over the whole area. The Amphibia are represented by the smooth newt, the frog and the common toad. The fish fauna of the lochs and rivers include the salmon, the sea and the common burn trout, the char, the pike, the perch and the eel. The marine fishes are so numerous that it would be impossible even to name them.

The invertebrate animals, both land and sea, include a vast assemblage of species. The Mollusca are very numerous, and there is not a rock between tide-marks that is not covered with barnacles. The sandy shore is alive with sandhoppers and the littoral region abounds in shore-crabs, shrimps and sea-slaters, while in the greater depths are to be found the edible crab and lobster. Star-fishes, sea-urchins and cucumbers are represented by a large number of species, the most abundant star-fish being the common cross-fish. Besides these there is a vast profusion of still lower forms, including the zoophytes, jelly-fishes, sea-anenomes and corals. The most abundant anemone is the beadlet (*Actinia mesembryanthemum*). The cup coral (*Caryophyllia*) occurs sparingly in Oban Bay. A considerable number of species of sponges are also found off the Argyllshire coast.

Among the land shells, *Helix nemoralis* is widely distributed, while other common species belong to the genera *Bulimus*, *Pupa* and *Clausilia*. The shire is also rich in the different orders of insects. The lakes and ponds abound in a great variety of Crustacea, Coelenterata and Protozoa.

## 9. Along the Coast.

It would be far from easy to perambulate in reality the two thousand odd miles of the Argyllshire coast line—a greater distance than from Ireland to Newfoundland.

Loch Long and Loch Goil

We begin our survey at the head of Loch Long, which extends for 17½ miles. In its upper reaches at Arrochar the scenery is of a wild Alpine grandeur, and the mountains dip sheer into the sea. As we descend the loch, they become less rugged and of lower altitudes. Near the mouth of Loch Goil, the western offshoot of Loch Long, both the 25 feet and 50 feet beach may be traced as rocky shelves along the Loch Long shore. As a rule, however, this feature is almost entirely absent. On the west shore of Loch Goil stand the ruins of Carrick Castle, an ancient stronghold of the Dunmore family. Proceeding down the loch we pass in succession Ardentinny, Blairmore and Strone, where the loch opens out into the Firth of Clyde. Rounding Strone Point we reach the Holy Loch, which measures only about two and a half miles in length, the north side flanked by Kilmun Hill. The various villages and piers on the shores of the loch are Kilmun, Ardnadam or Sandbank, and Hunter's Quay. Continuing round the Cowal peninsula, we pass Dunoon and then Inellan. Beyond Inellan is Toward Point, the termination of the Cowal peninsula, upon which the Toward Lighthouse, built by Robert Stevenson in 1812, is situated. Entering the Kyles of Bute, we have on our right the opening to Loch Striven, an arm of the sea which strikes northward for about eight miles; the shores of this loch are of a somewhat bleak and sombre character and there are but few habitations to be seen. Some distance further up the Kyles is the pier of Colintraive. The name, signifying the "swimming narrows," is derived from the fact that the Bute graziers used to swim their cattle across at the narrow part of the Kyles when returning from the Argyllshire markets. At this point the channel becomes greatly contracted, while the occurrence of several low-lying reef-like islands makes one imagine he is sailing into a veritable *cul de sac*. A short distance further on, Loch Riddon opens from the bend or elbow of the Kyles, penetrating Cowal for a distance of four miles to the mouth of Glendaruel. Near the mouth of the loch is a small island called Eilean Dearg, or red island, on which are the remains of a fort erected in 1685 in connection with the Earl of Argyll's expedition from the Netherlands. After rounding Rudha Ban Point, we reach Tighnabruaich, a pretty and sequestered watering place. Passing round the headland of Ardlamont, we enter Loch Fyne, extending over 50 miles in a north-easterly direction. The loch presents a great variety of shore and mountain scenery. In its upper reaches it is screened by such lofty mountains as Ben An Lochain and Ben Bheula. In the neighbourhood of Inveraray its shores are richly wooded, the town and the Castle being built on the raised beach platform which fringes the loch. The converging valleys of Glen Shira and Glen Aray also form conspicuous features in the landscape at this point. Lower down the loch the scenery gradually softens and the hills are not so wild in outline. Loch Gilp, which lies in line with the southward reach of Loch Fyne, has a length of three miles. On the west side is the village of Ardrishaig and on the opposite side is Kilmory Castle, and the graveyard and site of the ancient chapelry of Kilmory. At the head of the loch is the well-built little township of Lochgilphead near the entrance to the Crinan Canal. Six miles below Ardrishaig is Tarbert Bay or East Loch Tarbert which forms a fine land-locked natural harbour, long noted as one of the fishing ports for the far-famed Loch Fyne herring. In the neighbourhood is a picturesque old ruin, the remains, tradition says, of Robert the Bruce's Castle of Tarbert. Passing along the east side of the peninsula of Kintyre, which projects 42½ miles into the Atlantic and varies in width from 4½ to 11 miles, we reach Campbeltown Loch. Rounding the bold broad promontory, the Mull of Kintyre, we face the open waters of the Atlantic Ocean. From Machrihanish Bay to West Loch Tarbert the coast line is somewhat monotonous, but it presents some fine maritime views of the channels of Jura and Gigha, terminated by the long outline of Jura and Islay, in which the Paps form a predominant and beautiful feature. Loch Tarbert is a long inlet reaching so near to the smaller Loch Tarbert on the eastern side that the intervening neck of land is only about a mile in width. Further on, Loch Killisport and Loch Sween trench deep into Knapdale. The shores of the former are rich with rocks and wood, while those of the latter are indented in a remarkable manner. Towards the upper part the hills become abrupt and rocky, and are wooded from the water's edge high up along their acclivities. The scenery of this loch is striking and extraordinary. The coast line between Loch Sween and Loch Crinan is wild and rugged. At Loch Crinan is the western entrance of the canal, the other extremity of which is in Loch Gilp. On the northern shore of the loch stands Duntroon Castle, a conspicuous feature in the landscape where buildings of visible magnitude are rare. Loch Craignish now opens up, extending six miles inland. The great attraction of this loch is its cluster of islands, over twenty in number, as well as the profusion of islets and rocks, unnamed and uncounted. Perhaps the most remarkable of all the narrow passages in the Western Isles is that of the "Dorus Mor," or "Great Door," off Craignish Point, through which the tides run with a velocity of nearly eight miles an hour. It was here that the first "Comet" was wrecked in 1820, the strong wind proving too much for its feeble four-horse-power engine. Rounding Craignish Point we come in sight of Craignish Castle. The Tertiary dykes here often stand out like walls and castles, having a striking resemblance to architectural ruins. This is due to the much greater hardness of the dykes than the surrounding phyllites and their consequent superior resisting power to the action of the waves. Loch Melfort is mountainous in outline and is sprinkled with islets. To the north of Loch Melfort the mainland as far as the shores of Loch Etive and Loch Creran is mainly occupied by a series of volcanic escarpments with a general north-north-east and south-south-west trend, corresponding to the successive outcrops of the lava flows. These are well seen round Loch Feochan. On the left we have the islands of Luing, Seil and Shuna, which exhibit an extensive range of picturesque and pleasing scenery.

Proceeding up the narrowing Firth of Lorne, we pass Kerrera with Gylen Castle on the left and turn into the land-locked Bay of Oban. Leaving Oban, we see on the right Dunolly Castle picturesquely situated on a crag of Old Red conglomerate. By degrees Loch Etive opens up to the right, with Dunstaffnage Castle standing upon a wooded peninsula at the mouth and the peaks of Ben Cruachan in the distance. Loch Etive trenches into the heart of the county for about 20 miles. At Connel Ferry, near the mouth, the sides contract till the loch is barely 200 yards broad. The depth at low water is only six feet and the tides, which rise about 14 feet, rush with tremendous force through the narrow channel, breaking into angry foam, with a roar which may sometimes be heard miles off. On the left is the verdant island of Lismore, an example of the greenness of the vegetation over the limestones of the Highlands. From Loch Creran to Ballachulish the eastern shores of Loch Linnhe become steeper and its mountains rise in elevation. At Ballachulish the remarkably glaciated rocks along both sides of Loch Leven may be noted as well as the platform of the 50 feet raised beach, which projects from its northern side. At Corran Ferry Loch Linnhe appears as if cut in two by the same beach, whose level green surface makes a striking contrast with the dark, rough, mossy slopes on each side. At the top of the loch there is another expanse of the raised beach as well as moraine mounds, while straight in front lies the opening to the Great Glen. The peninsula of Morven is bounded by the western shore of Loch Linnhe, the Sound of Mull and Loch Sunart. Rounding Barony Point, we reach Ardtornish Castle, which Sir Walter Scott makes the gathering place of magnates and minstrels to do honour to the nuptials of the Maid of Lorn.

> "'Wake, Maid of Lorn!' the minstrels sung.
> Thy rugged halls, Artornish, rung,
> And the dark seas, thy towers that lave,
> Heaved on the beach a softer wave."

In passing through the Sound of Mull we have on either side the terraced basalt slopes of Mull and Morven, best seen as we approach Salen. Past the Sound we reach the mouth of Loch Sunart, which separates the peninsulas of Morven and Ardnamurchan. The loch has a length of 19½ miles, winding westward to within five miles of Loch Linnhe, and containing a number of islets. The peninsula of Ardnamurchan is the extreme northern part of the mainland of Argyll; and Ardnamurchan Point, the most westerly portion of the mainland of Scotland, is a wild bluff promontory, which on account of its savage and rocky character and the prevalence of the western swell is more terrible to mariners than any other headland between Cape Wrath and the Mull of Kintyre. For about ten miles eastward from the promontory much of the seaboard consists of well-cultivated arable land; but further to the east, where the volcanic rocks are replaced by the crystalline schists, the land is cultivated only in irregular and scanty patches. At the entrance to Loch Moidart we reach the limit of the Argyll coast line.

Passing now to the principal islands, we begin with Mull. The island is of an extremely irregular form, due to the many indentations of its coast line, which is estimated to be about 300 miles in length. The coast line along the Sound of Mull is of a comparatively unbroken character. The greatest irregularity is exhibited on the west and south coasts, especially the former. The principal inlets, taken in order from north to south, are Loch Cuan, Calgary Bay, Loch Tuadh between Ulva and Mull, Loch na Keal and Loch Scridan. From the west side of the island the marked contrast that exists between the flat cakes of basalt that form the plateau country of the northern portion of the island and the intrusive masses which form its south-eastern portion can be best seen. Almost every point on its shores is rocky or precipitous. The south-east coast exhibits considerable variety of cliff scenery. It rises to a mean altitude of over 2000 feet, culminating in Benmore 3169 feet above sea level.

The coast line of Coll and Tiree, islands composed of the Archaean gneiss, present an intermixture of rocky shores with small sandy bays. In Coll the rocks prevail and the coast line is generally bold, rising into bare rocky eminences having an altitude which nowhere exceeds 326 feet above sea level. Tiree is comparatively flat, being considered the lowest and flattest of the Hebrides.

Iona or Icolmkill is situated at the south-west corner of the Island of Mull and is separated from the Ross of Mull by a channel about a mile in width. The island consists of Archaean gneiss. Its coast line is for the most part indented with small rocky bays, divided by similar promontories. On the north-western side there is a large plain terminating in a flat shore of sand, largely made up of fragments of broken shells. A similar sandy plain on the east contains the ancient remains and modern village.

The upland consists of a mixture of rock and pasture of a moorish character. The great interest of the island is centred in its associations with St Columba.

Staffa, the name of which signifies "pillar-island," from the Scandinavian *stafr* and *ey*, is a small uninhabited isle off the west coast of Mull. The principal attraction of this celebrated island lies in the remarkable grotto,

**Fingal's Cave, Staffa**

Fingal's Cave, scooped out of an ancient lava stream, whose thick vertical columns rising from the level of the sea and divided by joints, form the sides of the cavern. The roof consists of smaller interlacing columns. The cavern has an architectural regularity of structure which bears some resemblance to a Gothic cathedral. The sea finds access to its floor of broken columns and during fine weather a boat may enter the cave. The solemn grandeur of Fingal's Cave makes it well worthy of being associated with the greatest of Ossian's heroes. The caverns of the island bear witness to the enormous erosive power of the ocean breakers.

> "Down-bearing with his whole Atlantic weight
> Of tide and tempest on the structure's base,
> And flashing to that structure's topmost height,
> Ocean has proved its strength, and of its grace
> In calms is conscious, finding for his freight
> Of softest music some responsive place."

The Island of Colonsay exhibits many points of beauty. The hills and cliffs do not reach any great altitude, yet they present a certain rugged grandeur which makes a pleasing contrast with the richness of its valleys and the yellow sands of its shores. Oronsay is separated from Colonsay by a sound which is dry at low water for a period of three hours. Raised beaches, the 25 feet, the 50, and the 100, are well represented on these islands, nearly every bay and headland showing evidence of a gravelly deposit or rock shelf.

The Island of Jura ranks third in size of the Argyllshire islands, and is celebrated for the height and shape of its mountains with their quartzite heads, the Paps of Jura. The island is one of the most rugged and bleak of the Hebrides. The western slopes are steep and rough, devoid of verdure and intersected by innumerable torrents. The eastern slopes are more gentle, and clothed with vegetation, presenting a more pleasing aspect to the eye. Between Jura and Scarba is the celebrated whirlpool of

Corrie-Bhreacain, the maelström of Celtic mariners and the subject of many tales and traditions.

The sea-board of Islay is generally bounded either by low rocks or flat shores and sandy beaches, but at the Mull of Oa it rises into great cliffs 656 feet in height. Immediately above the shore there is usually green pasture or arable land. Loch Indal opens on the south of the island between the Mull of Oa and the Point of Rhynns and penetrates twelve miles inland. The Rhynns of Islay lighthouse stands on the islet of Oversay at the west side of the entrance to the loch.

## 10. Climate and Rainfall.

The climate of a country is largely dependent upon its latitude, shape, exposure to the sea or to a particular point of the compass, its elevation above sea level, the character of its river and valley systems, nature of its soils, and the humidity and temperature of the air, the last two being perhaps the most important.

The cause of the motion of the air known as wind lies originally in change of temperature and consequent alteration of the weight of the air over different parts of the globe. The movements of the air may be either cyclonic or anticyclonic. Cyclones are areas of low barometric pressure with an encircling system of winds. Cyclonic systems usually bring to the region which they cover a large amount of cloud and rain and may be described as bad weather systems. The anticyclone on the other hand proceeds from an area of barometrical

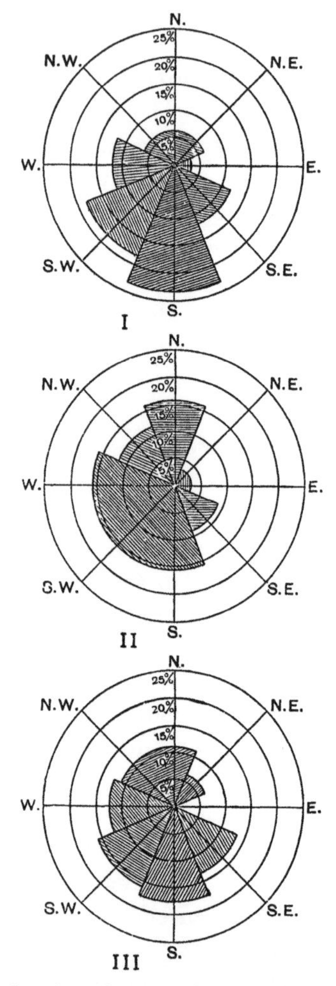


Diagram showing the prevalent winds at Skerryvore lighthouse in January, July, and throughout the year

I. January. II. July. III. For the whole year.

high pressure. It is less active than the cyclone and is characterised by gentle breezes blowing spirally outwards. The weather under its influence is generally quiet and dull. The winds of Scotland throughout the year are influenced by three more or less permanent pressure centres —a low pressure area south of Iceland ; a high pressure area situated in the Atlantic near the Azores ; and a continental area in Europe and West Asia, high in winter and low in summer. The Icelandic and continental areas are in prominence during the winter months and give rise to a great swirl between them, which causes the wind to blow from a south-west or north-east direction.

The direction of the prevailing winds at Skerryvore lighthouse, which is situated on an extensive reef 10 miles to the south-east of the Island of Tiree, is expressed in the diagrammatic form known as a wind rose (see figures on p. 46) and embody the average of observations taken over a considerable period. Along each of the eight principal points of the compass in these diagrams a distance has been marked off proportional to the percentage of days on which the wind blew in that direction. In the top diagram, which represents the winds for January, it will be seen that the prevailing winds are those from the south and south-west. The same holds good for the month of July, as appears in the second diagram. The third diagram shows that these are the prevalent directions of the wind for the whole year.

The prevailing winds, from the south and south-west, have the comparatively high temperature of the Atlantic ocean and, being at the same time surcharged with its vapour, they cover the high mountains with clouds, which discharge their contents in the shape of rain over the neighbourhood and keep the atmosphere in a humid state.

One of the principal points that has been deduced from a study of the rainfall of Scotland is the enormous difference between the west and the east. The stations along the west coast show such figures as 40, 45, and 54 inches as compared with 24, 27, and 30 inches on the east coast, unless in the immediate neighbourhood of the hills. Keeping in mind that the great source of rainfall is the prevailing south-westerly wind, we are enabled to show that the low fall in the east is due to the high land lying to the south-west, which robs the winds of a large proportion of their moisture in their passage across. On the other hand, the mountainous region of the west Highlands, deeply indented with arms of the sea which run in all directions from south to west, has currents of moist air continuously poured in upon it, with the result that this district has an enormously high rainfall. Thus at the Bridge of Orchy it amounts to 113·62 inches and at Upper Glencroe to 127·65 inches.

The following table shows the actual increase in rainfall as we proceed from the coast line towards the watershed:

| | |
|---|---|
| Islay (Ealabus) ... ... | 49·28 |
| Lochgilphead ... ... | 56·12 |
| Oban ... ... ... ... | 64·18 |
| Glen Fyne ... ... ... | 104·11 |
| Bridge of Orchy ... ... | 113·6 |
| Upper Glencroe ... ... | 127·65 |

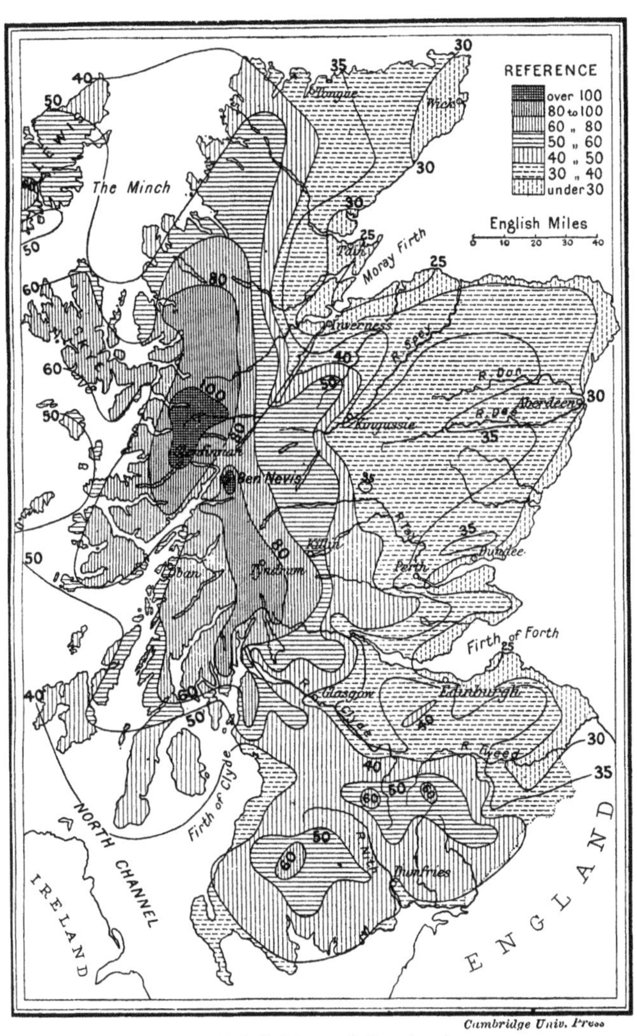


**Rainfall Map of Scotland**

*(By Andrew Watt, M.A.)*

In the neighbourhood of the sea snow seldom lies for more than two or three days, and this is true even for the shores of Loch Eil and Loch Linnhe, where they approach Ben Nevis, the highest mountain in Scotland. On the tops of the mountains, however, snow lies for four or five months of the year, and Ben Cruachan is seldom clear of it before the end of July. The winters are milder in Argyllshire than in many parts of the south of Scotland, and even of England. That portion of the county bordering on the Firth of Clyde has long been a favourite retreat for those suffering from pulmonary complaints; and it is stated by some medical men that the inhabitants of the Western Islands are almost entirely free from such complaints.

## 11. The People—Race, Language, Population.

In the work of the Alexandrian geographer, Ptolemy, who lived in the second century A.D., Kintyre is named Epidium and its inhabitants Epidii. Who these were by race and tongue, it is impossible to say.

Towards the end of the fifth century, we are on firmer ground, when a colony of Dalriad Scots, that is, Goidelic Celts, led by Fergus Mor and his two brothers, migrated from Ireland—only fourteen miles from the Mull of Kintyre—and settled in the region that is now Argyllshire.

Three centuries later came the Norsemen. Again

Iona Peasants

and again they ravaged island and mainland, made settlements, and intermarried with the Celts. The Gallgaels were this mixed people. Later admixture is comparatively slight, though we may note that during the "troubles" of the seventeenth century, Covenanting families—Hunters, Wallaces, Montgomeries—migrated from Ayrshire and

Cottage at Dalmally—Ben Cruachan in background

Renfrewshire to Kintyre. The ancestry of the present natives of Argyllshire must be mainly sought in the Scoto-Norse men of the Gallgael days. The physical characteristics of the Norse persist, Norse antiquities abound, Celtic customs and beliefs have been influenced by the Norse, the Norse tongue, though it died out as a distinct language, has left its traces on Gaelic. Nowhere is this influence so distinct as in the place-names, especially of the islands. It is calculated that of the island names ninety per cent. contain the Norse *ey*, "island," in the forms *-a*, *-ay*, or *aidh*, as in *Eriska*, *Ulva*, *Gigha*, *Staffa*, *Sanda*, *Jura*, *Oransay*, *Islay*. The Norse *ness*, "headland," appears as *nish* in *Craignish*, *Ardtornish*, *Treshnish*. Some of the names are wholly Norse, as *Ulva*, the first part of which is from *ulfr*, "wolf"; or *Jura*, the first part of which is from *dýr*, "deer." Other names, in whole or in part, of Norse origin are *Ardlamont*, *Melfort*, *Dunstaffnage*, *Skipness*, *Saddell*, *Carradale*, *Ormidale*, *Ealabus*.

In language, the Highland boundary line has continued as a sharp line of demarcation between Celtic and Teutonic. To the north of that line Gaelic is the vernacular tongue, to the south English is the universally spoken language. Argyll is one of the Scottish counties in which Gaelic is largely used. The last returns show that Gaelic speakers in Argyll number 31,695 and constitute 44·7 of the total population. In 1901 they numbered 40,577 or 52·1 per cent. of the total. Those now returned as speaking both Gaelic and English number 30,340, and as speaking Gaelic only 1355, the former being 6694, or 18·1 per cent., less than in 1901, and the latter 2168, or 61·8 per cent., less.

The population of the county (see also map on p. 135) is but sparsely distributed, and the last census shows a still further decrease. In 1801 the population was found to be 71,859, and the three following censuses showed an increasing population, the maximum being reached in 1831, when it totalled 100,973. From that period it shows a steady decline till it has now fallen to a little over 70,000, less than on all the previous census years. This decrease is attributable to various causes, the chief being emigration to foreign countries, the

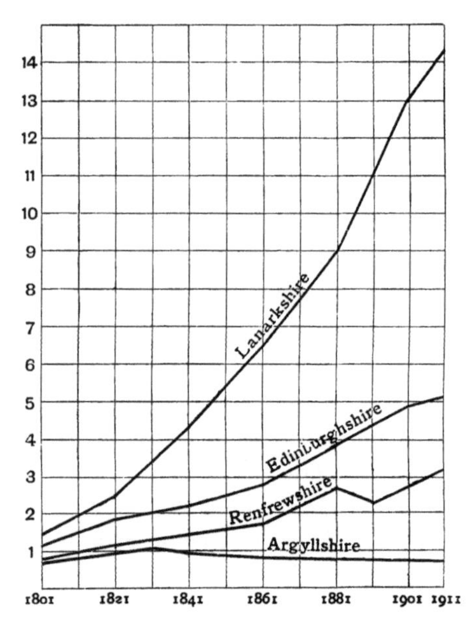


Population of Argyllshire, 1801–1911, compared with that of other Scottish counties

(*The numbers on the left-hand side represent population in hundred thousands*)

attractions of city life, and principally the concentration of the population in the mineral-producing counties of the Scottish Midlands.

## 12. Agriculture.

By far the larger part of Argyllshire is unfitted for agriculture, and up to comparatively recent times farming was in a backward condition. The abolition of the feudal system, the construction of the Crinan and Caledonian Canals, the promotion of a system of farming suited to the capabilities of the soil, the diffusion of knowledge bearing upon the cultivation of land and the introduction of steam navigation have been important factors in the development of the agriculture of the county. The results, however, have been more felt in the department of live stock than in that of husbandry. The chief crops are oats and hay. The former is raised principally for the ease with which it ripens under a minimum of sunshine. The lack of sunshine and the abundance of moisture account for the long time taken to dry the grass crop ; and the hay is often impoverished by rains before it can be stacked. The following figures from the Government Agricultural Statistics give the acreage devoted to the various crops during 1911 : hay 27,172 acres, oats 17,132 acres, barley 1436 acres, turnips and swedes 5792 acres, potatoes 3764 acres, peas 4 acres.

The Kyloes or West Highlanders have long been the staple cattle of the county. They are a small shaggy race and are considered much superior to the Dunrobins and Skibos or North Highlanders. Though small of size, they are greatly esteemed in the market and are exported in great numbers to Glasgow and other places on the Clyde.

Considerable numbers of Ayrshire cattle are reared in Islay, Kintyre, and the southern parts of the county, and also in the neighbourhood of the towns and villages where milk is in demand.

The sheep belong to the black-faced breed. They were first introduced into the south of Scotland from Northumberland and afterwards into Argyllshire. Of a

Highland Cattle

hardy type, they are well suited to the country. Their mutton is good but their fleece is somewhat coarse. The horses of Argyllshire have been crossed with other breeds, especially the Clydesdale, and have lost all distinctive character. A few ponies may still be picked up in Mull and Tiree. A large number of pigs are reared in the county for the Glasgow market.

At one time the shire, like the rest of the Highlands, must assuredly have been covered by dense forests, and the following passage is probably not overdrawn: "All these islands and western shores once waved with giant trees that would rival the American backwoods. The impenetrable forest of Calydon extended all over Argyll,

Caledonian Forest near Inveroran

its terrible depths peopled by wild bulls, boars, and bears, and wilder Britons formed an impassable barrier even to the invincible legions of Rome. In the peat mosses which cover so large an extent of the western isles roots of forest trees in great quantities are found in the position in which they grew 5 or 6 feet beneath the surface of the super-accumulated moss. On a steep rocky bank by the house (Ardlussa, Jura, in 1861) stands a most venerable witness to this fact in the presence of a hollow-hearted old oak tree 21 feet in circumference though very dwarfed in height. A great deal is dead but some boughs yet have leaves. Edinburgh savants opine that this tree is more than 1500 years old. Another smaller one, a mere boy which has probably not yet seen 1000 summers, stands near."

The treeless condition of Mull in 1773, when Dr Johnson visited it, moved him to jibes. His oak-stick had disappeared and nothing could persuade him that it had not been stolen. "No, no, my friend," said he, "it is not to be expected that any man in Mull who has got it, will part with it. Consider, sir, the value of such a *piece of timber* here!" When Sir Allan McClean talked of the woods of Mull, Johnson remarked, "Sir, I saw at Tobermory what they called a wood, which I unluckily took for *heath*. If you shew me what I shall take for *furze* it will be something."

## 13. Industries and Manufactures.

The manufactures of Argyllshire are not important. Formerly a large quantity of kelp used to be gathered along the shores for the extraction of iodine, but this industry is now extinct. Between 1725 and 1730, an important ironwork was established by Irish capitalists at the mouth of the Kinglass on the east side of Loch Etive, at which pig-iron and castings were both made. About the same period ironworks also existed at Bonawe and Innerleckan near Inveraray. They manufactured pig-iron from haematite brought from Ulverston and other localities on the north-west coast of England. The products were taken back to England. The Innerleckan works continued down to 1813, those at Bonawe till much more

Campbeltown

recent times. Similar ironworks also existed in Islay. The reason for the establishment of ironworks in these localities was that amid the general treelessness they possessed abundance of timber, which was used as fuel in the smelting of the iron. It is possible, however, that some of the thicker veins of ferriferous carbonate found in the neighbourhood may have been used for smelting. Coarse woollen yarns, stuffs and stockings are still extensively made in certain districts. At one time an attempt was made to introduce the weaving of book muslin into Islay, weavers having been brought from Glasgow, but it was not successful. In the same island the spinning of yarn was at one time a staple industry, as much as £10,000 worth being exported in a year, but it was unable to compete with the Glasgow manufactories. In Islay and at Campbeltown the distillation of whisky is extensively carried on.

The Campbeltown Shipbuilding Company was founded in that town in the year 1877 and first devoted its attention to the construction of wooden vessels. During the past 33 years the Company has constructed about 100 steamers of all types for owners in almost every part of the world. The yard is situated on the north side of Campbeltown Loch, about one mile from the town, and is equipped with all the most modern shipbuilding plant and labour-saving appliances. In normal times it gives employment to a large number of men, all of whom belong to the district and have been trained locally by the firm.

## 14. Mines and Minerals.

Auriferous ore has been recently extracted from a small lode near Stronchullin, Knapdale. The gold, which is invisible, is associated with lead, copper and zinc ore, in a matrix of quartz. This appears to be the only authentic record of the occurrence of gold in Argyllshire.

Galena, the sulphide of lead, has been found in Islay at Mulreesh, where a large mine was formerly worked. At one time this seems to have been a profitable undertaking, but owing to the fall in the price of lead, accompanied by increased cost of production, the mine had to be abandoned. An assay of the silver lead from this locality showed 6½ ounces to the ton. At a more recent period Robolts Mine in the same locality was worked, but the ore was not found in sufficient quantities to be profitable. Many years ago manganese was mined in the south of the Oa promontory, where it occurred as strings or veins through the quartzite, but owing to the difficulty in procuring the ore the work was given up.

At Strontian, a village in Ardnamurchan parish, lead was worked from the beginning of the eighteenth century till 1855. The village gives its name to the element strontium, discovered in the strontianite. The mineral occurs as fibrous green or brown masses. Other minerals found in the locality are sphalerite and fluorite.

The Ben Cruachan granite has been quarried for a considerable period at Bonawe. The granite is of a fine-grained type and makes an excellent building stone. It seems, however, to be principally used for the making of granite setts for road paving; and is extensively shipped to Glasgow and other places. The granite is also crushed for the manufacture of cement. Further up the loch at the Craig quarries a coarser type of granite is obtained, which can be extracted in large blocks. It has been much used for building and ornamental purposes and for harbour work on the Clyde. Granite has also been wrought at various places on the Ross of Mull. The mass of quartz porphyry lying between Furnace and Crarae on the shores of Loch Fyne has been largely quarried. It is extensively used in Glasgow for the making of pavement setts.

Inveraray Castle and other buildings in the neighbourhood have been built of a talcose schist from St Catherine's and Creggan. This stone is homogeneous in structure, dark-green in colour, and somewhat soapy to the touch. Easily wrought it can be used for rough carving while it offers a greater resistance to the weather than any of the other stones in the district. An inspection of the stone used in Inveraray Castle shows the original chisel marks still in a wonderful state of preservation.

A bed of coal belonging to the Carboniferous system has long been worked in the neighbourhood of Campbeltown. It is of no great extent but it is thick and of good quality. Coal of Tertiary formation occurs in Mull, but in such small quantities and in such an inaccessible position as not to pay the working.

There are several slate quarries in the county. The oldest, those at Easdale, were commenced about the year 1631 and they are still carried on vigorously. During 1906 these quarries gave employment to over 350 persons, an important means of livelihood in what is otherwise a sparsely populated pastoral district. Roofing-slate has also been extensively quarried at Ballachulish, where the outcrop commences at the shore and stretches south and eastwards along the side of the mountain. The following statistics of the production of roofing-slate in Argyllshire

Ballachulish: the slate quarries

is quoted from the "General Reports and Statistics on Mines and Quarries." The fluctuation in the output is due largely to the labour troubles at Ballachulish.

| *Year* | *Quantity in Tons* | *Value in £* |
|---|---|---|
| 1900 | 25,713 | 51,947 |
| 1901 | 26,446 | 51,805 |
| 1902 | 27,531 | 54,371 |
| 1903 | 24,439 | 48,309 |
| 1904 | 32,336 | 59,247 |
| 1905 | 17,786 | 38,141 |
| 1906 | 24,323 | 45,303 |

Limestone is more or less abundant in several districts but the largest area is in the island of Lismore. The lime is used chiefly for manuring by the local farmers, and to a less extent for building purposes. Formerly it was extensively worked, but now the demand is insufficient to give constant employment throughout the whole year. Quartzite is quarried at Port Appin and conveyed to Glasgow, where it is ground for use in the manufacture of delf ware.

## 15. Fisheries and Fishing Stations.

Loch Fyne has long been celebrated for the quantity and quality of its herring. The herring-season lasts from June to January; but white fishing is carried on all the year round. The mode of fishing for herring is by drift-nets. Another mode of fishing for herring chiefly practised in Loch Fyne is known as "trawling," but the instrument of capture is really a "seine" net. The exceptionally fine quality of the Loch Fyne herring is generally attributed to the temperature and to the superior feeding found in that loch. The actual records of net fishing make it very difficult to account for the increase or decrease of fish in such waters as Loch Fyne. Formerly it was supposed that the shoals of herring

Fishing fleet on Loch Fyne

had been driven from our shores by steamships and trawlers, and yet many of our recent seasons have broken all previous records. Some naturalists maintain that the herrings never leave Loch Fyne at all, but when they appear to leave the loch they really go to the bottom to feed in quiet, afterwards returning to the warmer inshore waters in order to spawn. Two of the 27 fishery districts of Scotland have their headquarters at Inveraray and Campbeltown. In the latter district during 1911 102,848 cwt. of fish were landed, valued at £26,375. These were taken by 318 boats, and the men and boys employed numbered 735. The total value of the boats and fishing gear was estimated at £20,371. The principal kinds of fish caught were herring, cod, saithe, haddock, whiting and plaice.

Not only is the herring fishery of great importance in itself, but the movements of the herring have considerable influence over those fish which prey upon the fry. The principal of these are the cod, whiting, haddock, mackerel and flat fish. These are all found from time to time in the various lochs, and are also known to frequent particular banks. The fishery for crabs and lobsters in the Clyde estuary is now unimportant, being confined to the south of Arran and the Kintyre coast.

No part of the west Highland coast is richer in salmon than the rivers and lochs of this region. The fish are as a rule both finely shaped and well fed, though varying according as they come from the Shiel, the Aline, the rivers bordering Loch Linnhe, those of Mull, those at the head of Killisport, and Loch Indal in Islay. It is difficult to obtain particulars as to the quantities of fish captured throughout the district, but the total must be very large. Loch Shiel and the river Shiel are famous for their salmon. It is a peculiar though well-known fact that certain waters are untenanted by the salmon, while in others they will not rise to the fly. According to certain authorities the migratory salmon on their way from the deep sea first strike the coast at Lochbuie and then proceed up Loch Linnhe to the many fine waters to the south of Loch Linnhe. Those for the north coast and the Lochy take the Sound of Mull.

## 16. History of the County.

The Dalriad Scots, who settled in Argyllshire about the year 500, were destined to be the source of two important developments in Scottish history. From this region Christianity, through the labours of St Columba and his disciples, reached the Picts and the Northern English; and from a Dalriad king sprang the dynasty that ultimately ruled over a united Scotland.

At first Dalriada comprised the mainland of modern Argyllshire with Islay and Jura, its capital being Dunadd on Loch Crinan. For some time the settlers had a hard struggle with the Picts, who under King Brude hemmed them into Kintyre. But better days arrived after Columba landed in Iona in 563—perhaps the greatest date in the history of Argyllshire. Columba was no mere ecclesiastic, he was also a soldier and a statesman. He was instrumental in consolidating the Dalriad kingdom under King Aidan and in securing it against extinction by the Picts. For the next three centuries the kingdom had many ups and downs in its conflicts with the neighbouring nations. With the coming of the Northmen, the Vikings, in the eighth century, whose attacks made the Western Islands insecure, the religious centre for Scots and Picts was transferred from Iona to Dunkeld, in the reign of Constantin I.

In 844 the King of Dalriada, Kenneth MacAlpin, became ruler of the united Picts and Scots, who were never again separated as nations. Norse invasions still continuing, Kenneth followed the example of Constantin

Farm Steading, Iona

and conveyed relics of St Columba to Dunkeld. The invaders did more than merely raid : they began to found settlements in the western isles and on the coast of the mainland. About 890, after Harold Harfagr had made himself master of Norway, he led a great expedition against the Vikings of this region, subdued them and gave the Hebrides—Sudreys or South Islands, as the Norse called them—to Ketil Flatnose. The islands remained Norwegian for several centuries, but by no means submissive to the kings of Norway or friendly neighbours to the kings of Scotland. In 1102 Magnus Barefoot came, like Harold Harfagr, and again subdued the Sudreys. He bargained with King Edgar of Scotland, claiming as Norwegian all the western islands "between which and the mainland a helm-carrying ship could pass." Then Magnus played Edgar a pretty trick. Seated in his galley, he had it dragged across the isthmus between East Loch Tarbert and West Loch Tarbert, and thus secured the peninsula of Kintyre.

When Malcolm IV came to the throne in 1153, Somerled—whatever his race, his name is Norse, "the summer-sailor"—King of Argyll, rebelled against his nominal head, the Scottish king, and warred for three years. In 1164 Somerled again attacked Malcolm, but along with his son he was assassinated on the coast of Renfrew. In 1222 Alexander II subdued Argyll and parcelled it out among his followers. Some forty years later the battle of Largs ended the Norwegian claim over the Western Islands. But till the seventeenth century Argyll and its islands were again and again scenes of turbulence. Individual families, as the MacDougalls of Lorne were contumacious and fought against their superior; the Lord of the Isles repeatedly sought to become independent and intrigued with the enemies of the King of Scotland. Several of the Jameses went in person against their western foes, or entrusted the task of crushing them to some powerful magnate, as the Earl of Argyll—representative of the great and illustrious family which began to be prominent in the thirteenth century in the person of Sir Colin Campbell of Lochow or Loch Awe.

In 1609 an important gathering of chiefs took place in Iona to meet representatives of the king. The chiefs had to accept the "Band and Statutes of Icolmkill," which marks a new departure in the discipline of the Highlands and Islands. Next year the chiefs became responsible for the good behaviour of their clansmen. Four other notable incidents of this century belong to Argyllshire. In December 1644 Montrose pierced the snow-covered passes and appeared at Inveraray, plundering the lands of his great rival Argyll.

In 1651–52 English troops surrounded Inveraray and finally forced Argyll to come to terms with Cromwell. The next head of the family returned in 1685 from exile in Holland to his ancestral domains to raise a rebellion in favour of the Duke of Monmouth, and failed. In 1692 a wild February night saw the grim tragedy, the too well-known Massacre of Glencoe, "the Glen of Weeping."

It remains merely to note in the eighteenth century the Appin murder, a mysterious affair, when near Ballachulish a Campbell was shot dead in 1762. Allan Breck Stewart was suspected but evaded capture. Another Stewart, a supposed accomplice, was tried at Inveraray, and condemned by a jury of Campbells. Need the reader be reminded of the graphic account of this incident in R. L. Stevenson's *Kidnapped*?

Inveraray

## 17. Antiquities.

Archaeologists divide the prehistoric period of man's occupation of a region into the Ages of Stone, Bronze, and Iron according to the material of which his weapons of war and the chase or his implements of industry were manufactured. The Stone Age is subdivided into the Palaeolithic or Old Stone Age, and the Neolithic or New Stone Age. In the former the weapons were rudely chipped into shape; in the latter they were smooth, polished, and better adapted to the end in view.

The antiquities of Argyllshire are numerous and varied, including hammer heads, celts, axes, knives, and other remains of the Stone Age. A number of years ago a most interesting discovery of several shallow caves or rock shelters was made at Oban. They were situated in the cliff which bounds the esplanade, and had no doubt been formed by the action of the sea at the period of the formation of the 25—30 feet beach. The bottom of one of the caves was covered by a layer of pebbles and gravel, thus indicating that the sea had had access to it. In these caves were found the remains of men, women and children, and the bones of *Bos longifrons*, red and roe deer, pig, goat, badger and otter, shells of edible molluscs, bones of fish and claws of crab. In association with these were found flint scrapers, hammer stones, and implements of bone and horn, which had been made into the form of pins, borers and chisel-shaped implements. Prof. Turner, who described these relics along with Dr Anderson,

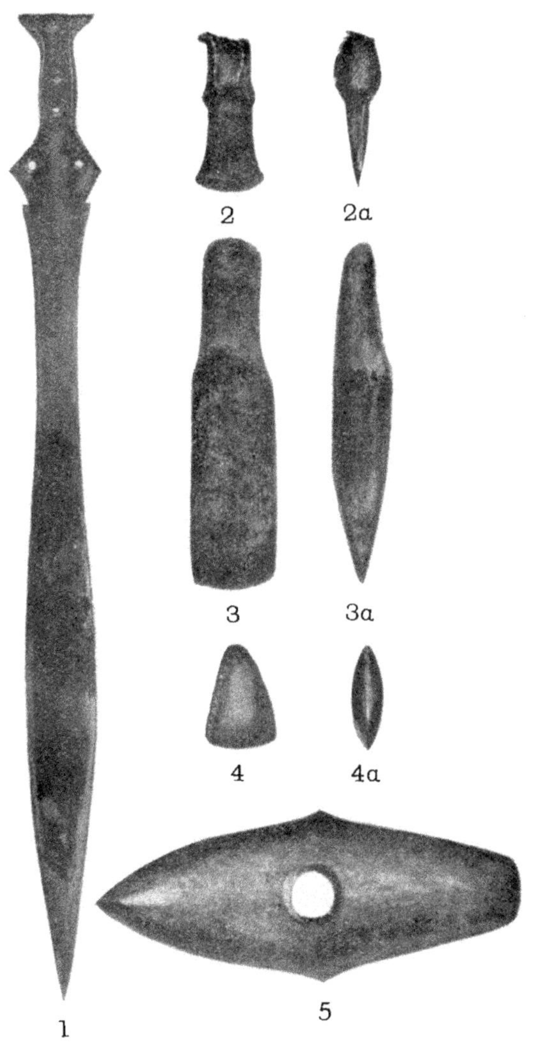


**Stone and Bronze Implements from Argyllshire**

1. Bronze Sword, Island of Shuna. 2, 2*a*. Bronze flanged Axe Head, Island of Mull. 3, 3*a*. Axe of Diorite, Glen Fruin, Helensburgh. 4, 4*a*. Axe of Diorite, Tarbert. 5. Axe Hammer of Granite, Inveraray

remarks that in one of these caves several harpoons or fish spears made of the horns of deer were obtained, these being comparable with similar implements which have been found in England and France associated with Palaeolithic objects.

The relics of the Bronze Age are abundant throughout the shire and include the following: a leaf-shaped sword from Shuna; winged and socketed celts from Mull, Barcaldine, North Knapdale, Southend, Kintyre; knife daggers from Cleich, Loch Nell, and Callachy, Island of Mull; socketed knife and spear head from Campbeltown; and a chisel-like instrument from Strachur. Stone moulds in which these implements were cast have also been found at Campbeltown and Kilmartin. Throughout the country there occur many stone circles, some of them in a wonderfully perfect state of preservation. These are generally known as Druid Temples, but they have no claim to this name. They are now supposed to have been associated with the burial customs of the Bronze Age.

Vestiges of vitrified forts are to be found at various places in the county. Perhaps the most notable of these is that mistakenly named Beregonium, on the east of Ardmucknish Bay, not far from Connel Ferry. Tradition makes it the site of the ancient capital of Dalriada and the Selma of Ossian. Nothing remains but a well-defined fort, in some parts eight feet high, on the top of a hill, and fragments of a defensive wall at the base.

The ancient Celtic crosses that are distributed throughout the mainland and islands of Argyll were introduced into this country by the missionaries from Ireland. In the Irish type the cross-bearing slabs are generally recumbent. In Argyllshire and the Western Isles the developments are in the form of standing crosses, a good example of which is the cross at Kildalton. Besides zoomorphic and spiral ornamentation, this cross bears representations of the Virgin and Child, the sacrifice of Isaac, and David rending the lion's jaw. It measures over eight feet in height and over four in width across the arms.

Kildalton Cross

Many interesting relics of the Norsemen have been found—brooches and other ornaments in a Viking burial place in Oronsay; sword, axe, shield-boss, cauldron, a pair of scales, coins of the ninth century in another Viking burial place at Killoran Bay, Colonsay. Brooches of Norse origin also come from Islay and Tiree. A cross-shaft of the Viking Age was discovered in St Oran's Chapel in Iona, and has been deposited in the Cathedral of Iona. It shows the Scandinavian dragon, a galley with its crew, a smith with hammer, anvil and pincers.

At Ballachulish what seems to be an idol of the pagan Norsemen, perhaps the divinity from a Viking ship, was dug from peat in 1880. The image, a female figure nearly life-size, is of oak and has quartz pebbles for eyeballs.

The MacDougalls of Dunolly, Oban, have long possessed the famous Brooch of Lorne, which fell into the hands of their ancestor, the Lord of Lorne, after the battle of Dalry in Perthshire. The brooch fastened the mantle of Robert the Bruce, who when hard pressed by three of Lorne's vassals killed them but could not free the brooch from the grasp of one of the dead men. The brooch consists of a silver plate about four inches in diameter, with a rim notched to look like battlements. Eight pearl-crowned cones rise within the rim. In the centre is an unpolished gem resting on elegantly carved work. For the story of the Bruce's loss of the brooch, see Scott's *Tales of a Grandfather* and *Lord of the Isles*.

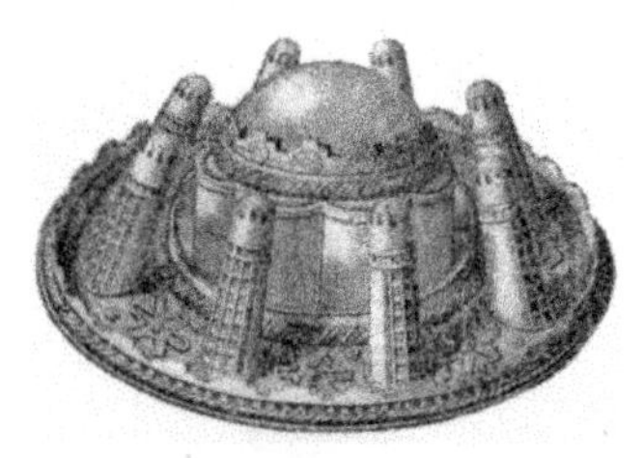

The Brooch of Lorne

## 18. Architecture—(*a*) Ecclesiastical.

In every parish, in every island, and in many of the islets of Argyllshire the ruins of early Christian oratories or churches are to be found. These are identified principally because they were built of stone, and a doorway, a window, an altar, a piscina, or a cross gives evidence of their ecclesiastical purpose. The original Celtic buildings were much more primitive, being formed of such easily procured materials as wattles, wood or turf for the walls, and bracken or heather for the thatch.

In 563 Columba and his disciples founded the monastery on the island afterwards known as Icolmkill, that is, "island of Columba of the church. The modern name Iona is a mistake for Ioua, an adjective form used by Adamnan. The monastery consisted of a simple church of wattles and clay, huts of the same materials, and an earthern rampart for defence. This was the humble beginning of that religious foundation which from the sixth century to the eighth was hardly second to any monastery in the British Isles. It was then, as Dr Johnson said, "the luminary of the Caledonian regions, whence savage clans and roving barbarians derived the benefits of knowledge and the blessings of religion." But events, as we have seen, made Iona unsafe. Norse raiders continually descended on the island, plundering and murdering. On Christmas Eve, 986, for example, they ravaged the island and slew the abbot with fifteen of the monks. Time and the violence of men destroyed all the Columban buildings and their immediate successors. The oldest existing structure is St Oran's Chapel, with a fine Norman doorway and triple arch. The Nunnery Church is Transitional, with marked characteristics of the Norman style tending towards the pointed. Iona Cathedral is usually classed in the third pointed period because the greater part of it belongs to a late date. It is built of the red granite of the Ross of Mull and various kinds of rocks; and is roofed with heavy slabs of mica slate. It consisted of nave, transepts and choir, with sacristy on the north side of choir, side chapels on the south, and a tower in the

Iona Cathedral and St Martin's Cross

usual position. The length ot the cathedral is about 160 feet. The tower, 70 feet high, is lighted on one side by a plain slab pierced by quatrefoils and on the other by a beautiful Catherine wheel or marigold. The carvings upon the columns and tombs are sharp and well preserved.

A visit to the island makes one feel what Dr Johnson

expressed in the words: "That man is little to be envied, whose patriotism would not gain force upon the plain of Marathon, or whose piety would not grow warmer among the ruins of Iona."

The only other cathedral in Argyllshire was the cathedral on the Island of Lismore. Erected early in the thirteenth century, it was in the first pointed style, 137 feet in length and 39 in width. All that now remains is a rectangular chamber, which has been used as the parish church. At Kildalton in Islay there is a church, of the first pointed style, 60 feet long, with a canopied piscina.

## 19. Architecture—(*b*) Castellated.

The ancient castles of Argyllshire, now mostly in ruins, are very numerous and have a general uniformity of style. They usually exhibit a high square tower surmounting a rocky cliff or other eminence, many of them overhanging some stream or the sea. These old towers were often the abode of an almost incredibly large number of inmates, and they tell of the habits of people inured to war. They were sparingly lighted, and they must have been both gloomy and unwholesome. They often exhibit curious tracery, and a surprising profusion of inscriptions, armorial bearings and miscellaneous devices.

Carrick Castle, an ancient stronghold of the Dunmore family, on the west side of Loch Goil, dates from the end of the fifteenth century, and, though roofless, is a good specimen of mediaeval strength. At West Tarbert on Loch Fyne are the ruins of Tarbert Castle, once a residence of King Robert the Bruce. Most of what remains of this castle is fifteenth and sixteenth century work. Duntroon Castle, an ancient baronial fortalice on a promontory projecting into Loch Crinan, was originally surrounded by a wall some six feet thick and

**Dunolly Castle**

from 24 to 28 feet high with a walk round the top and an embattled parapet. The old castle has been repaired and modernised into a comfortable mansion. On the south point of the Island of Kerrera are the ruins of Gylen Castle, dating from the twelfth century. Long a stronghold of the MacDougalls of Lorne, it has been described as one of the most charming little specimens of architecture of Scotland before the national art was absorbed in the general European style of the Renaissance.

Dunolly Castle, the ancient stronghold of the Lords of Lorne, stands on a cliff on the northern horn of Oban Bay. Little now remains of the original structure but the ivy-clad donjon. Fragments of other buildings which surround it attest that at one time it must have been a place of considerable importance.

Two miles to the south-west of Connel Ferry are the spacious ruins of the ancient and venerable Dunstaffnage Castle, on a bold promontory jutting into Loch Etive. It appears to have been a square building having round towers at three of its corners. The present entrance is by a ruinous staircase to the sea and is supposed to have been protected by a fosse and drawbridge.

Castle Stalker occupies a rocky islet in Loch Linnhe off the mouth of Appin Bay. It is an ancient square tower, roofless, but fairly complete. It was built by Duncan Stewart of Appin and was used as a hunting seat by his relative King James IV. Over the entrance gate is a fine carving of the royal arms.

Duart Castle, the ancient seat of the Macleans, rises from a bold headland north of Achnacraig on the Island of Mull. It is supposed to be of Danish construction, and comprises a massive square tower 75 × 72 feet.

Ardtornish Castle was a great stronghold of the Lords of the Isles. It is situated on a low basaltic headland at the entrance to Loch Aline—

> "Artornish on her frowning steep
> 'Twixt cloud and ocean hung."

Little now remains of its former glory but a square keep and some fragments of outer defences. It was from this castle that the Lord of the Isles in 1461 granted in the style of an independent sovereign a commission to two of his kinsmen to confer with the deputies of King Edward IV concerning an attack upon Scotland.

Mingarry Castle was twice occupied by James IV, 1493 and 1495. It stands on the south shore of the Ardnamurchan peninsula at the mouth of Loch Sunart. The castle is an irregular hexagon in outline, and is defended by a dry ditch. It is surrounded by a high wall, forming a kind of polygon for the purpose of adapting itself to the projecting angles of the precipice overhanging the sea. The principal entrance was in the south wall, convenient for access to ships and for communication among the islands.

Kilchurn Castle, one of the most magnificent and picturesque ruins in the Highlands, occupies a promontory at the east end of Loch Awe, and is an example of an earlier keep enlarged into a castle surrounding a courtyard. The five-storied keep was built in 1440 by the famous Sir Colin Campbell, Knight of Rhodes, an ancestor of the Breadalbane family. The castle forms an oblong quadrangle with one corner truncated and each of the others flanked by round hanging turrets.

Carnassary Castle, on an eminence at the head of Kilmartin valley, was built by John Carswell, first rector of Kilmartin and Bishop of the Isles after the Reformation: he translated Knox's liturgy into Gaelic and had it printed 1567—the first book printed in that language.

Kilchurn Castle, Loch Awe

The castle is in the style of an ancient keep, and the internal arrangements are such that if the castle was attacked the keep could be shut off from the other apartments and used as a place of safety. On the ground floor is the kitchen, with great arched fireplace and oven. The first floor has a common hall and private room. A parapet runs round the top, and the staircase is surmounted by a cape.

Ardchonell Castle, now a picturesque ivy-mantled ruin on Loch Awe, was in the sixteenth century a stronghold of the ancestors of the Duke of Argyll. Towards the end of the fifteenth century the infant heir to the Lordship of the Isles, as a prisoner of great importance, was carefully guarded in the castle. In this locality originated the Campbells' famous slogan, "It's a far cry to Lochow."

The ruins of Toward Castle opposite Rothesay have not been inhabited since the attempt of the Campbells in 1646 to exterminate the Clan Lamont. The most interesting architectural feature of the castle is the ornamentation of the arch of the entrance gateway, which is a beautiful example of the revived early work of the sixteenth century. Near at hand is the new castle.

## 20. Architecture—(*c*) Municipal and Domestic.

As the towns within the county are but small, municipal architecture is not on a very large scale. The principal edifices are the Dunoon Burgh Buildings in

Inveraray Castle and Duniquoich

the Scottish Baronial style, the Inveraray County Court House and Prison, and the Court House at Oban.

Only a few of the mansions can be here described. Inveraray Castle, the principal seat of the Argyll family, stands on the right bank of the Aray, near the town of Inveraray. It is beautifully situated on the raised beach platform fringing Loch Fyne, while behind it rises

Duniquoich with its richly wooded sides and an ancient watch tower perched on its bald crown. The castle was built by the third duke, 1744–61. It is in the Gothic of the eighteenth century, and has a bold and imposing appearance with towers at the angles. The park, upwards of thirty miles in circumference, is nobly wooded and highly picturesque, with three splendid avenues, one of limes and two of beeches. Dundarave Castle, a principal seat of the MacNaughtens, is situated on the shores of Loch Fyne, 4½ miles above Inveraray. The castle, which consisted of a strong tower with turrets at each angle, has recently been subjected to a restoration rather meretricious than classical. On the opposite shore of the loch at the foot of Glen Kinglass stands in beautiful grounds the mansion house of Ardkinglas, built of the local quartz porphyry. It succeeded a house destroyed by fire about 1840.

Toward Castle, at the eastern entrance to the Kyles of Bute, is a fine Gothic mansion, built in 1821. Craignish Castle is situated on Craignish peninsula. Included in the modern mansion is a strong old fortalice which withstood a six weeks' siege by Colkitto. Armaddie Castle, a seat of the Marquis of Breadalbane, pleasantly situated at the head of a small bay opposite Seil Island, commands to the south-west a magnificent vista of sea and shore. The building, which is very old, belonged to the MacDougalls, Lords of Lorne, but afterwards passed into the possession of the Argyll Campbells.

## 21. Communications.

Compared with its population, no district in Great Britain enjoys such facilities for steamboat traffic as the county of Argyll. The West Highlands generally have benefited greatly by the modern improvements in steamboat communication. In the early part of last century the want of means of transit was severely felt, and the old Statistical Account contains many references to this grievance. The first steamer built in this country, the old "Comet," made her trial trip from Glasgow to Greenock on the 18th January, 1812, and in September of the same year her voyage was extended to Oban and Fort William. Passenger boats were placed on the Crinan Canal in 1838, and this made a large increase in the traffic. Luxuriously appointed steamers now receive the passengers at either end of the canal. The service from the Clyde to Inveraray, Oban, Skye and the Western Isles, initiated by Messrs Burns of the Cunard line, was taken over about the year 1851 by Messrs David Hutcheson and Co., who developed the trade with great vigour. It is now in the hands of Messrs David MacBrayne, Ltd., who have a fine fleet of passenger and trading steamers known as the "Royal Route" to Oban.

The Crinan Canal lies wholly within the county, and the Caledonian Canal has its south-western end in Argyllshire. The former was constructed, 1793–1801, under the auspices of John, Duke of Argyll, at a cost of £142,000.

It has 15 locks, an average depth of 10 feet, a surface width of 66 feet and bottom width of 30 feet. Vessels of 200 tons can pass through it. The canal runs across the isthmus from Ardrishaig to Loch Crinan and enables vessels to avoid the circuitous route round the Mull of Kintyre.

The Tyndrum to Oban section of the Caledonian

A lock on the Crinan Canal

Railway, with the extension to Ballachulish, lies within the shire, and a small portion of the West Highland Railway line passes through the extreme east of it, north and south of Bridge of Orchy. Campbeltown and Machrihanish are connected by light railway.

The roads usually take advantage of the deep glens, and as the watersheds separating the glens are often low, they rarely ascend to great heights. There are, however, some exceptions. Thus the road between Glendaruel and Otter Ferry reaches the height of 1026 feet; that between Lochgoilhead and St Catherine's 727 feet; and that between Glencroe and Cairndow 860 feet. The last is on the main road between Glasgow and Oban *via* Inveraray. The road up Glencroe was formed by one of the regiments under General Wade after the rebellion of 1715. For the distance of a mile and a half the road ascends in a zig-zag fashion, and to the average pedestrian is difficult and fatiguing. A stone inscribed "Rest and be thankful" is placed at the summit. Wordsworth, who traversed the road on one of his tours in Scotland, wrote

> "Doubling and doubling with laborious walk,
> Who that has gained at length the wished-for height,
> This brief, this simple wayside call can slight,
> And rests not thankful?"

## 22. Administration and Divisions.

The Royal Burghs of the county are Inveraray and Campbeltown. A Royal Burgh, created by a charter from the Crown, has the right of self-government by a magistracy and council, and possesses many important privileges. Inveraray was made a Royal Burgh in 1648 and Campbeltown in 1700. The county returns a member to parliament. Inveraray, Campbeltown, and Oban belong to the Ayr district group of parliamentary burghs.

The Lord-Lieutenant is the Duke of Argyll, with thirty-one deputy-lieutenants, of whom at present (1913) no fewer than twelve are Campbells. For several centuries the head of the Campbells was hereditary sheriff, and in 1747, when hereditable jurisdictions were abolished, Argyll received £20,000 in compensation. Argyllshire has now a sheriff-principal, with three sheriffs-substitute at Dunoon, Campbeltown, and Oban. Courts are held also at Inveraray, Lochgilphead, Tobermory, and Bowmore in Islay.

The most important administrative body in the shire is the County Council, which looks after the finances, roads, bridges, public health and general administration. The districts are: Ardnamurchan on the Atlantic; Morven bounded by Loch Sunart, the Sound of Mull and Loch Linnhe; Appin on Loch Linnhe; Benderloch lying between Loch Creran and Loch Etive; Lorne surrounding Loch Etive; Argyll in the middle of the shire; Cowal between Loch Fyne and the Firth of Clyde; Knapdale between the Sound of Jura and Loch Fyne; and the long peninsular district of Kintyre, stretching southwards from Lochs Tarbert to the Irish Sea.

There are 28 civil parishes in the mainland of Argyll: Ardchattan and Muckairn, Ardgour, Ardnamurchan, Campbeltown, Craignish, Dunoon and Kilmun, Glassary, Glenorchy and Inishail, Inveraray, Inverchaolin, Kilbrandon and Kilchattan, Kilcalmonell, Kilchrenan and Dalavich, Kilfinan, Killean and Kilchenzie, Kilmartin, Kilmodan, Kilmore and Kilbride, Kilninver and Kilmelford, Lismore and Appin, Lochgoilhead and Kilmorich,

Oban

Morven, North Knapdale, Saddell and Skipness, Southend, South Knapdale, Strachur, Stralachlan. The insular parishes are: Coll; Colonsay and Oronsay; Gigha and Cara; Jura; in the island of Islay, Kilarrow and Kilmeny; Kilchoman; Kildalton; in the island of Mull, Kilfinichen and Kilvickeon; Kilninian and Kilmore; Pennygown and Torosay; Tiree.

The ecclesiastical parishes make up seven presbyteries (one of them, Abertarff, includes some Inverness-shire parishes) in the synod of Argyll.

The educational matters of the county are under the jurisdiction of the School Boards. There are secondary schools at Campbeltown, Dunoon and Oban. Tarbert public school has a secondary department and several other schools earn grants for giving higher education.

## 23. Roll of Honour.

Among the famous men connected with Argyllshire, by birth or residence, we have already mentioned Fergus Mor, who led the Dalriad Scots; Aidan, their great king in the sixth century; St Columba; his biographer Adamnan; Kenneth MacAlpin, who united the Picts and the Scots; and Somerled, "probably," says *The Clan Donald*, "the greatest hero that his race has produced."

The members of the illustrious house of Argyll cannot be all noted here. The Campbell family has by some been traced to Anglo-Norman origin as though its name were *de Campo Bello*. The late Duke of Argyll

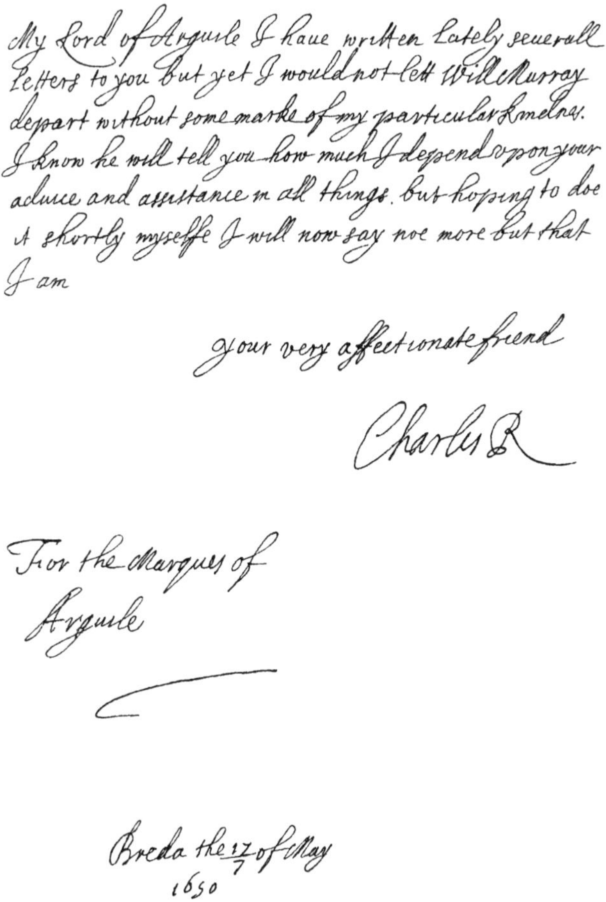

My Lord of Argyle I haue written lately severall
Letters to you but yet I would not lett Will Murray
depart without some marke of my particular kindnes.
I know he will tell you how much I depend upon your
aduice and assistance in all things. but hoping to doe
it shortly myselfe I will now say noe more but that
I am

Your very affectionate friend

Charles R

For the Marques of
Argyle

Breda the 17/7 of May
1650

Letter from Charles II to the First Marquis of Argyll

took it to be purely Celtic, of Scoto-Irish origin. The regular old form is *Cambel*, which is simply *cam beul*, "curved mouth." Archibald, eighth Earl and first Marquis of Argyll, born in 1607, was an eminent statesman and patriot. Succeeding his father in 1638, he joined the Covenanters in opposing Charles I. In 1644 he took the field against Huntly, and dispersed the royalists, but he was no match for Montrose, who penetrated into Argyllshire in December of the same year, and in the beginning of 1645 annihilated Argyll's men at Inverlochy. Opposed to the execution of Charles I, Argyll placed the crown of Scotland on Prince Charles's head at Scone. Finally, he complied with Cromwell's government, and for this he was tried after the Restoration. He was beheaded, 27th May, 1661.

His son, the ninth Earl, distinguished himself at the battle of Dunbar and continued to hold out against the Protector's forces till Charles commanded him to submit. After his father's execution he was received into the royal favour. In 1681 he took the Test Act, with a reservation, for which he was sentenced to death on the charge of "leasing-making." He escaped to Holland, whence he returned in 1685 to aid in the rebellion against King James. Taken prisoner at a ford on the Cart in Renfrewshire, he was executed in Edinburgh on his former sentence.

John Campbell, second duke of Argyll, was one of Marlborough's generals and distinguished himself at Ramillies, Oudenarde and Malplaquet. He did much to promote the Act of Union. Born, as Pope says, to "shake alike the senate and the field," he was a renowned statesman, played a decisive part in placing George I on the throne, commanded the royal troops at Sheriffmuir, and took a firm stand against the proposals to punish Edinburgh after the rioting of the Porteous Mob.

George John Douglas Campbell, eighth duke of Argyll, born in 1823, was a famous statesman and writer. He held office in several ministries from 1853 to 1881. His writings include *The Reign of Law*, *A History of the Antiquities of Iona*, *Scotland as it Was and as it Is*.

To the legal profession Argyll has given at least one distinguished son, Lord Colonsay, born in the island of that name in 1794. He adorned the highest legal offices in Scotland, and was the first Scotsman to be transferred from the supreme Scottish Court to the House of Peers.

Many eminent divines are associated with the county. Dr George Campbell, born 1706, became professor of Church History in St Andrews. Dr John McLeod Campbell, born at Armaddy House near Kilninver in 1800, was ordained to the charge of Row parish, Dumbartonshire, in 1825. He was deposed for teaching doctrines contrary to the faith of the Church. He then commenced a fresh ministerial career in Glasgow, where he continued to preach till 1859. Dr Norman Macleod, one of the most attractive preachers and writers connected with the Church of Scotland in recent times, was born in Campbeltown in the year 1812, and ultimately became minister of the Barony Church, Glasgow. In 1860 he undertook the editorship of *Good Words*, which he managed with extraordinary success. Among his more important writings are *Reminiscences of a Highland Parish*, *The Old Lieutenant and his Son* and *The Starling*.

Argyll cannot boast of any great names in science; but Colin Maclaurin, the eminent mathematician, was born in the parish of Kilmodan in 1698. Educated at Glasgow University, he became professor of mathematics in Marischal College, Aberdeen, in 1717, and in Edinburgh University in 1725. A controversy with Bishop Berkeley led to the publication in 1742 of his greatest work, the *Treatise on Fluxions*. He also wrote *Geometrica Organica*, and a treatise on the tides, which shared the prize offered by the French Academy. He died in 1746.

The great folklorist, John Francis Campbell, was born in Islay, 1822. He was a profound Gaelic scholar, and his *Popular Tales of the West Highlands* "is one of the most important contributions ever made by any scholar to the scientific study of folk-tales." He also invented the sunshine-recorder.

Of the Gaelic poets of the county, the foremost is Duncan Ban Macintyre, or the fairhaired poet, who was born in Glenorchy in 1724. In his early life he was employed as a forester on the Breadalbane estate. His poem on *Beinn Doireann* is considered one of the finest examples of modern Gaelic poetry.

## 24. THE CHIEF TOWNS AND VILLAGES OF ARGYLLSHIRE.

(The figures in brackets after each name give the population in 1911, and those at the end of each section are references to pages in the text.)

**Ardrishaig** (1297), Gaelic *ard-driseach*, "height full of briers," is a seaport village in South Knapdale parish. Situated on the west side of Loch Gilp at the entrance to the Crinan Canal, it is the centre of an extensive herring fishery and shares in the Canal trade, especially in shipping cattle and sheep. (p. 37.)

**Ballachulish** (2781), Gaelic "town of the strait," a straggling village on the southern shore of Loch Leven on both sides of the Laroch river. There is a ferry across Loch Leven. (pp. 40, 62, 76, 88.)

**Bowmore** (805) is a small seaport town, with a safe anchorage, standing on the east side of Loch Indal, in Islay. It is the capital of the island and has a distillery. (p. 90.)

**Campbeltown** (7625) is a thriving town situated upon Campbeltown Loch, defended by the island of Davarr and affording excellent harbourage. The principal trade used to be herring fishing and the loch is still the rendezvous of hundreds of fishing smacks and wherries as well as larger and more important vessels. The manufacture of whisky is perhaps the most distinctive feature of the place. Other branches of industry are a large net manufactory, a flourishing shipbuilding yard, carpenter's yard, a rope manufactory, and wrights' and engineers' shops.

The Drumlemble colliery is in the neighbourhood. Most of the houses are situated at the head of the loch, hence the Celtic name *Ceann-loch*. In 1700 it was called Campbeltown out of compliment to the Duke of Argyll. (pp. 60, 66, 74, 89, 92, 95.)

**Dunoon** (6859), one of the most frequented watering places on the western shore of the Firth of Clyde, extends for more than three miles and may be regarded as practically one with Kirn and

Dunoon

Hunter's Quay. Dunoon proper is the part fringing the east and west bays. The lower part of the town stands upon the raised beach, while the higher part straggles up the hillside. Behind rise the heather-clad Cowal Hills, culminating in Bishop's Seat. (pp. 20, 26, 85, 92.)

**Easdale** (216) is a small village situated on the island Easdale about sixteen miles to the south-west of Oban, off the island of Seil. On Seil and opposite Easdale is another small village, Ellanabriech. The strait between the islands is the regular highway for the villagers, who are principally employed in the slate quarries. (p. 62.)

**Inveraray** (533), a seaport town and a royal and parliamentary burgh, on Loch Fyne about six miles from its head. The town itself is not particularly attractive but this is made up for by the great beauty of the surrounding scenery, especially along the banks of the Aray and the Shira. The town was founded in 1742 in place of an earlier settlement, which stood in front of the Castle. At the foot of the street is a sculptured stone cross with a Latin inscription, which is now almost illegible. (pp. 11, 37, 59, 62, 66, 70, 85, 89.)

**Kilmun** (1391) lies on the north-east shore of the Holy Loch about eight miles by road from Dunoon. Towards the close of the sixth century a Columban church was founded here by St Fintan Munnu. The ancient cell was replaced by a collegiate church built by Sir Duncan Campbell of Loch Awe in 1442. This has long formed the burying place of the Argyll family. (p. 36.)

**Kirn** is a fashionable watering place between Dunoon and Hunter's Quay. During the yachting season these places are astir with a large number of visitors.

**Lochgilphead** (921), at the head of Loch Gilp, near the Crinan Canal, is a well-built township, whose rise from a small fishing hamlet is due to the growing trade of the West Highlands. (pp. 38, 48, 90.)

**Oban** (5557), on a fine bay, near the opening of Loch Linnhe, is in the track of coasting vessels from north to south and is admirably adapted for trade. It is also the great rendezvous for tourists in the West Highlands. Near the end of the eighteenth century the Duke of Argyll, to encourage building, offered very favourable terms to those desirous of erecting houses. The town owes much of its prosperity to two brothers Stevenson, who settled there in 1778. The town contains many large and splendidly appointed hotels while the hillsides are studded with villas, which are let during the summer season. (pp. 16, 21, 27, 40, 72, 76, 81, 85, 87, 89, 92.)

**Port Charlotte** (548), a small village on the west coast of Loch Indal in the Island of Islay.

**Port Ellen** (741), a seaport village in Kildalton parish, Islay, situated at the head of a small bay six miles north-east of the Mull of Oa. It was founded in 1844 and named in compliment to Lady Ellinor Campbell of Islay. It has become a favourite resort for golfers as the links are held to be unequalled, even in Scotland.

**Portnahaven** (478), a cod-fishing station, is situated at the south-western extremity of Islay at Rhynns Point. The island of Oversay protects the entrance to the bay from the fierce surges of the Atlantic, and has a lighthouse, erected in 1825.

**Sandbank** and **Ardnadam** (997), on the south side of the Holy Loch, are two of the summer resorts fringing the shores of the Firth and the mouth of the Holy Loch. Ardnadam consists of modern villas, but Sandbank is a much older village. (p. 36.)

**Strone**, at the headland of Strone Point in Kilmun Parish, extends along the north shore of the Holy Loch. This summer resort commands fine views of the Holy Loch and the Firth of Clyde. (p. 36.)

**Tarbert** (708), delightfully situated on East Loch Tarbert, is noted as a herring fishing port. Till recent years it was simply a village of fishermen, but it is now a favourite summer resort. (pp. 38, 80.)

**Tighnabruaich** (745), picturesquely situated and well-sheltered on the west side of the Kyles of Bute, has become of late years very popular as a watering place. To the south-west is the straggling village of Kames, with powder-works. (p. 37.)

**Tobermory** (988) is the capital of Mull and a modern seaport, on the north-east coast of the island. Its picturesque situation and its beautiful land-locked harbour lend an interest to the somewhat lonely town. It was founded in 1788 by the British Fishery Company but it never became a fishing station of any importance. Relics of the "Florida," a ship of the Spanish Armada blown up in 1588 by Maclean of Duart in the harbour of Tobermory, have been occasionally recovered. (p. 90.)

# BUTESHIRE

## 1. County and Shire[1]. Origin and Administration of Buteshire.

The islands comprised in the county of Bute seem in very early times to have belonged to different kingdoms —the Cumbraes to Strathclyde, Bute and Arran to Dalriada. After the Norwegian claim to this region was finally disposed of in 1266, the islands became the possession of the High Stewards of Scotland; and Bute was a sheriffdom or shire in the fourteenth century, for the Exchequer Rolls mention John Stewart as sheriff of Bute in 1388; and the same records speak of Arran and the Cumbraes as in the shire of Bute.

The name *Bute*, which appears in various spellings as *Boyet*, *Boet*, *Bot*, is of disputed origin. Some hold it to be the Irish *both*, "cell," in allusion to St Brandan's Cell erected in the sixth century. Others seek its origin in the British *buth* (Gaelic *bhioa*), "corn," since the island is fertile compared with others.

Buteshire has a Lord-Lieutenant, a vice-lieutenant, and six deputy-lieutenants; a sheriff—whom it shares with Renfrewshire—and a sheriff-substitute, resident in Rothesay. Sheriff courts are also held at Brodick and

[1] See p. 1.

Millport. The County Council consists of sixteen elected members with four representatives from the Town Council of Rothesay. Before the Reform Bill of 1832, Rothesay as a Royal Burgh returned a member to parliament, but is now merged in the county. From 1707 to 1832 Buteshire sent a member to parliament alternately with Caithness. In 1831 the county of Bute with a population of 14,000 had twenty-one electors, of whom only one resided in the county. At an election shortly before that date, nobody (besides the sheriff and the returning officer) attended except this one elector. He constituted the meeting, called over the roll of freeholders, answered to his own name, took the vote as to the Praeses, and elected himself. He then moved and seconded his own nomination as M.P., put the question to the meeting and was unanimously returned.

The educational matters of the county are under the jurisdiction of the school boards, and there is a secondary school in Rothesay.

The parishes are Cumbrae; Kilbride, and Kilmory in Arran; Kingarth, North Bute and Rothesay in Bute. Ecclesiastically these are separated. Cumbrae is in the Presbytery of Greenock under the Synod of Glasgow and Ayr. The parishes of Bute are in the Presbytery of Dunoon, those of Arran in that of Kintyre, under the Synod of Argyll.

## 2. General Characteristics. Natural Conditions. Communications.

Buteshire consists of the islands of Bute, Arran, Great and Little Cumbrae, Holy Isle, Pladda and Inchmarnock. It is wholly an insular county, the different islands being separated from each other and also from the mainland by the waters of the Firth of Clyde. In Arran especially both the inland and coast scenery is exceedingly fine. The great variety of picturesque beauty in the island and its accessible situation have made it a much prized retreat of the inhabitants of Glasgow. The island contains within its compass rugged Alpine-like mountains, swelling hills and open valleys. The shores are no less striking, exhibiting as they do nearly every variety of maritime scenery, here rising into bold cliffs and there sinking into open bays, which are diversified by woods and tracts of cultivated ground, with here and there a farm house or the remains of some ancient castle. The island of Bute presents features similar to Arran, though not on such a grand scale, as the granite mountains are absent from the northern area of Bute. Carlyle gives a pen picture of the island in his *John Sterling*: "The climate of Bute is rainy, soft of temperature, with skies of unusual depth and brilliancy, while the weather is fair. In that soft rainy climate, on that wild-wooded rocky coast, with its gnarled mountains and green silent valleys, with its seething rainstorms and many sounding seas, was young Sterling ushered into his first schooling in this world."

Millport, Cumbrae

The great Highland boundary fault passes through the islands of Arran and Bute, the southern halves belonging to the Lowland portion of the Scottish mainland, and the northern halves to its Highland portion. This has been an important factor in determining their outstanding geological and geographical features. It has been estimated that about one-third of the whole county is unprofitable and little more than one-sixth is under cultivation.

The west side and the north end of Arran are kept in communication with the mainland by the steamers which ply between Greenock and Campbeltown. Its east side has steam-boat connections with Greenock by way of Rothesay and Millport, and with Ardrossan in connection with trains for Glasgow. The south end of the island is visited by steamers running between Ayr and Campbeltown. Bute and the Great Cumbrae have abundant steam-boat connection with Craigendoran, Greenock, Wemyss Bay and Largs. A tramway connects Rothesay and Etterick Bay. A good carriage road runs round Arran, having a length of 56 miles. Two other roads intersect the island from shore to shore, from Shiskine to Brodick and from Lagg to Lamlash. Good roads traverse most parts of the island of Bute, and there is a road round the Great Cumbrae.

## 3. Size. Shape. Boundaries.

The County of Bute, measured from the northern extremity of the Island of Bute to the Island of Pladda, is $35\frac{1}{2}$ miles in length and from the north-eastern extremity of the Great Cumbrae to the western side of the island of Inchmarnock $9\frac{3}{4}$ miles in breadth. From the south-

**Pladda Island, from Kildonan**

western point of the Holy Isle in Arran to Drumadoon Point it has a breadth of $11\frac{1}{3}$ miles. The total area is 139,432 acres or 225 square miles.

Arran may be roughly described as elliptical, having its longer axis in a north and south direction and its minor axis in an east and west direction. Its greatest length is 19⅓ miles and its greatest breadth 10⅜ miles. It is bounded on the south-west and north-west by the waters of Kilbrannan Sound, which separates it from Kintyre in Argyllshire, on the north-east by the Sound of Bute, and on the east by the Firth of Clyde, which separates it from the Ayrshire coast.

The Island of Bute has a length of 15½ miles between Buttock Point in the north and Garroch Head in the south, with an average breadth of 3½ miles. It is bounded on the north, north-east and north-west by the narrow winding channel of the Kyles of Bute, which separates it from the Cowal District of Argyllshire, and on its south-easterly and south-westerly parts by the Firth of Clyde.

The Great Cumbrae, close to the Ayrshire coast, has been compared to a pointed tooth in outline, having Farland Point and Portachur Point in the south as fangs. About 3½ miles in length and two in breadth, it has a circumference of about 10½ miles with an area of 3120½ acres. It is surrounded by the waters of the Firth of Clyde. The "Wee" or Little Cumbrae, which lies 1½ miles south of the "Big" Cumbrae, is 1¾ miles in length and 7¾ furlongs in breadth, with an area of about 723 acres.

## 4. Surface and General Features.

The physical features of the southern and northern portion of the Island of Arran are exceedingly dissimilar in character. The northern half is a region of lofty and rugged mountains, mostly of granite. Towards the summit they are either destitute of vegetation or invested with a slight covering of Alpine plants and mosses. These great mountain masses are separated from each other by deep gorges and wild glens diverging from a common centre and radiating outwards on to the 25 foot raised beach. Some of the principal heights in the northern part of the island are: Goat Fell (2866 feet), Caisteal Abhail (2735), Ben Tarsuin (2706), north top of Goat Fell (2628), Cir Mhor (2618), Ben Nuis (2597), Beinn Bharrain (2345), Cioch na h-Oighe (2168), Am Binnein over High Corrie (2172), Ben Chliabhain west of Glen Rosa (2141), Meall nan Damh west of Glen Catacol (1870), Sail (1465), Chalmadale.

Few mountains in Scotland afford such a magnificent prospect as the summit of Goat Fell—an all-round view except on the west, where other lofty peaks intervene. Away to the north-west rise the Paps of Jura. Northwards the eye follows Loch Fyne to the Crinan Canal, with Ben Cruachan a little to the east, and possibly Ben Nevis in the far distance. More to the east are the Cobbler and Ben Vorlich, by the heads of Loch Long and Loch Lomond. Glancing over the south of Bute, we behold the gleaming waters of the Firth between Cowal and Renfrewshire, and over Dumbartonshire towers Ben Lomond. Turning east, we sweep the Ayrshire coast with its towns and towers and rising ground behind. To the south Ailsa Craig stands out, backed by the peninsula of Wigtownshire, while in the south-west we see Kintyre and remoter Antrim. In

Panorama of the Granite Mountains of Arran, from Ben Chliabhain

favourable weather we may even catch a glimpse of the Isle of Man.

The southern portion of Arran consists of an undulating tableland rising to a general elevation of from 500 to 800 feet, traversed by a series of ridges having a general east and west direction. The inland scenery of this portion is somewhat bleak and monotonous, but the coast reveals a great wealth of shore and cliff scenery of a beautiful and romantic character, especially where the cliffs advance upon the sea-line in mural precipices.

The highest points in this division are: Ard Bheinn (1676 feet), Cnoc Dhu north branch of Glen Cloy (1341), A'Chruach head of south branch (1679), Cnoc na Croise, head of Clachan and Benlister glens (1346), Beinn Bhreac, north-west of this (1649), Tigh-vein, south-west of Urie Loch, highest point south of the parallel of Lamlash (1497).

The Island of Bute is traversed by three valleys, or depressions, which run north-east and south-west and terminate on either side in bays or indentations of the land. The northernmost and largest runs from Kames Bay to Etterick Bay, the middle from Rothesay Bay to Scalpsie Bay, and the most southerly from Kilchattan Bay to Stravanan Bay. The principal elevations are Torran Turach (745 feet), Muclich Hill (638), North Hill of Bullochreg (769), Kilbride Hill (836), Kames Hill (875), and Eenan Hill (538).

The Great Cumbrae is similar to that portion of Bute north of Kilchattan and rises to about the same height. The interior is hilly, the greatest elevation being 417 feet. The Little Cumbrae, rising to 409 feet above the sea, is rocky like the southern end of Bute. Instead of marked ridges, however, there is a series of gently sloping terraces.

## 5. Watersheds. Rivers. Lakes.

The great central granitic area of north Arran is divided into two nearly equal parts by the glens of the Iorsa Water and Easan Biorach, which meet at the watershed of Loch na Davie. The former stream, about eight miles long and the largest in the island, flows south-west to the sea at Dougrie, draining the eastern slopes of Beinn Bharrain and Beinn Bhreac and the western slopes of Caisteal Abhail, Cir Mhor and others. The Easan Biorach runs north, entering the sea at Loch Ranza. From Caisteal Abhail and Cir Mhor a number of ridges diverge, and form the glens of Sannox and North Sannox, draining towards the east, and Glen Rosa, which drain southwards into Brodick Bay. On the north-west side of the granite mass the waters of Glen Catacol enter the sea at Catacol Bay. Some of the principal streams on the southern half of the island are those draining Glen Cloy, Benlister Glen and Monamore Glen, Kilmory Water, Sliddery Water, Black Water and Machrie Water. The only stream of any size in the Island of Bute is the Glenmore Burn, rising on Muclich Hill in the north and flowing south to Etterick Bay.

The surface of Bute and of Arran is diversified by a number of freshwater lochs. Nearly all those in Arran occur in the west. The largest, Loch Tanna, about a mile in length, is situated on the east side of Ben Bharrain, 1065 feet above sea level. It is probably not of any great depth and appears to be drift-dammed. To the west of it lies the Dubh Loch, which may be a rock basin. Loch Nuis, Loch Iorsa and Loch na Davie are shallow lochs probably in drift. The most picturesque of the

Loch Fad, Bute

Arran lochs is Loch Corrie an Lochain, between Meall nan Damh and Meall Biorach. It stands 1080 feet above sea level, surrounded by an amphitheatre of granite crags. In Bute Loch Fad and Loch Quien lie in the valley running from Rothesay Bay to Scalpsie Bay. Loch Ascog is about half a mile to the east of Loch Fad.

## 6. Geology[1] and Soil.

Arran has been classic ground for the geologist ever since the days of Hutton, the great founder of physical geology. Within its narrow compass it presents such a wide variety of phenomena that it has been described in a geological sense as an epitome of the British Isles. In recent years it has been even more carefully examined; and, though the claims of its more ardent admirers might be disputed, one cannot hesitate to affirm that it contains a fuller record of certain of the great Scottish formations than any other district of equal extent. In considering the geology of the county, it must not be forgotten that these islands are essentially a portion of the surrounding mainland, and what has been said regarding the geology of Argyll also applies to Bute. Thus we may note that the great Highland boundary fault enters the east side of Arran near Dougrie and circles round the granitic mountains, entering the sea between Loch Ranza and the Cock of Arran. From this point it crosses the Sound of Bute to Scalpsie Bay and proceeds by way of Loch Quien and Loch Fad to Rothesay Bay, crossing to Toward Point in the mainland. To the north-west of this line lie the crystalline schists of the Highlands, to the south the younger members of the Palaeozoic division, and in Arran sandstones of Triassic Age. The crystalline schists of the Highlands are best seen in the Island of Bute, and a glance at the geological map will show that

[1] See p. 15.

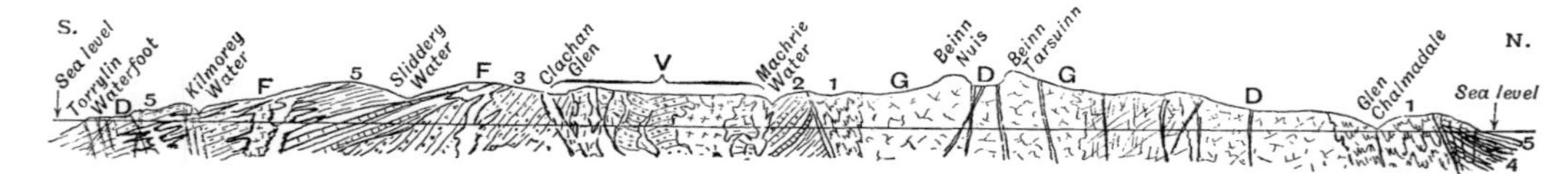


Fig. 1.

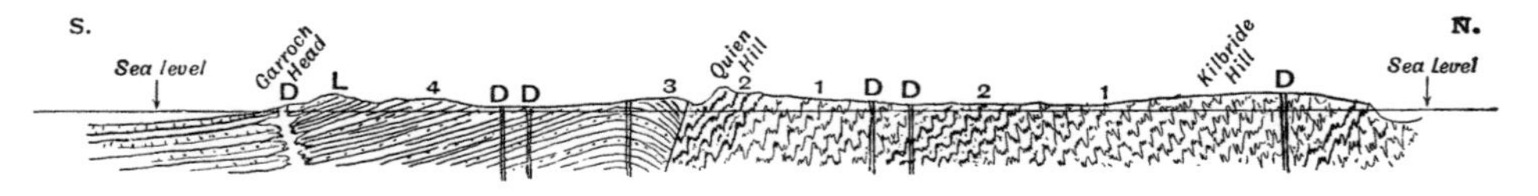


Fig. 2.

**Fig. 1. Section across Arran from Torrylin to the Cock of Arran**

1. Crystalline Schists. 2. Lower Old Red Sandstone. 3. Upper Old Red Sandstone. 4. Carboniferous. 5. Trias. D. Dykes. F. Felsite. G. Granite. V. Volcanic Neck.

**Fig. 2. Section across Bute from Garroch Head to Buttock Point**

1. Schistose Grits. 2. Phyllites. 3. Upper Old Red Sandstone. 4. Carboniferous. D. Dykes and Sills. L. Lavas.

they are simply the south-westerly prolongation of the bands seen on the Cowal shore, described on p. 20. The Dunoon phyllites and the schistose grits and grey-wackes which lie to the north of them are represented in the north of Bute. In the Island of Arran the crystalline schists form an incomplete ring round the granite of the north end. In the district between Glenshant Hill and Whitefarland these rocks form an elevated plateau rising to over 1000 feet. They also occupy the whole of the shore line from Loch Ranza to Dougrie.

In Glen Sannox there is a belt of black shales, cherts and grits with associated volcanic rocks similar in character to rocks found on the mainland at Aberfoyle. Quite recently, 1912, fossils were discovered in these rocks at Aberfoyle, clearly demonstrating that they belong to upper Cambrian times. This is of great interest to geologists as it not only shows that the Cambrian formation must have spread over the Highlands but also proves that the great earth movements which raised the Highland mountains was prolonged into Cambrian times.

The Lower Old Red Sandstone is separated from the schists by a fault, which can be traced from Corloch on the east of Arran to Dougrie on the west. Many of the prominent hills in the two Sannox Glens consist of conglomerates belonging to this formation. It rises to 1346 feet in Cnoc na Croise. On the west side of the island this formation contains a bed of lava, which marks the second great volcanic period in the history of the island. A large portion of Bute and the main part of the Great Cumbrae consist of rocks of Upper Old Red Sandstone Age, which are separated from the Highland

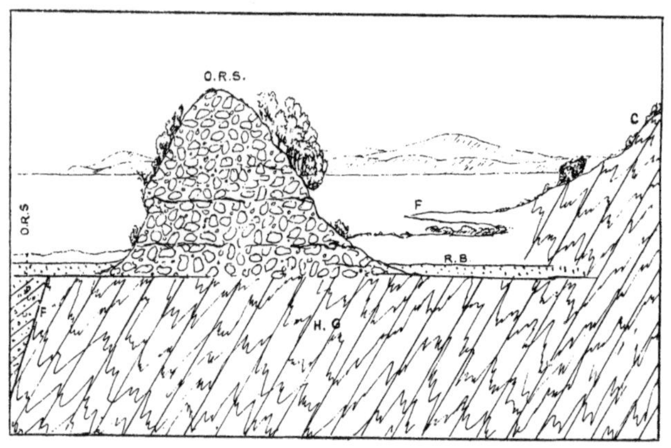


**The Haystack, Scalpsie Bay, Bute—a relic of denudation**

R.B. = 25 ft. Raised Beach. C. = Cliff behind Raised Beach. O.R.S = Old Red Sandstone. H.G. = Highland Grits. F. = Great Boundary Fault

schists by the great boundary fault just referred to. On the north side of Scalpsie Bay, Bute, there is a conical hillock called the Haystack, composed of coarse conglomerate which forms the basement portion of the Old Red Sandstone and rests unconformably on the schists. At one time a vast thickness of Old Red Sandstone and carboniferous rocks must have been piled upon it, but they have all been removed by denudation. These formations are thrown down to the south by the great boundary fault, which passes through the bay at this point, and have so been preserved. Behind the Haystack rises the cliff of the 25 feet beach which can be traced round the bay. Rocks of similar age appear in Arran between the Fallen Rocks and Corrie. They include the third phase of volcanic action on the island. Carboniferous rocks occur in Bute, in the Cumbraes, and in Arran. In Arran they can be seen along the shore between the Cock and the Fallen Rocks as a narrow strip cut off on the west by a fault. They roll over and are repeated to the south by an anticlinal arch.

The New Red Sandstone occupies the whole eastern shore of Arran southward from Corrie, stretching inland as far west as Glen Ormidale and Benlister Glen. On the west of the island it spreads over the vale of Shiskine on either side of the mouth of the Machrie Water.

In Arran fossils of Rhaetic, Liassic, and Cretaceous Age have been preserved in a somewhat remarkable fashion in the neck of an ancient volcano. The rocks which constitute the neck consist partly of fragmental and partly of crystalline igneous material. But in addition to these the vent contains fragments derived from the surrounding rocks, and also fragments of rocks not now found *in situ* in the island. These fragments are of considerable size and resemble the rocks of Rhaetic, Liassic and Cretaceous Ages found along the Antrim coast in Ireland.

Intrusive rocks of Carboniferous Age are numerous in the Great Cumbrae; and intrusive rocks of Tertiary Age form most of the outstanding features in the topography of Arran, including the great granitic mass of the Highland portion of the island.

## 7. Geology and Scenery.

As has already been pointed out, the northern half of Arran and Bute lies within the Highland region and the southern half within the Lowland region. The contrast between the two areas is especially well marked in Arran. To the north of the Highland boundary line tower the granite peaks with their surrounding belt of schist. The fine-grained granite and the coarse-grained granite produce hills of a type peculiar to each rock. This has been ascribed to the different manner of weathering of the two rocks. The deep nicks that form such a prominent feature against the sky line are generally due to the more rapid weathering of dykes intrusive in the granite. Ceum na Caillich is a good example of this phenomenon. A'Chir, the most serrated mountain in the island, is also traversed by a large number of dykes. The difference of weathering between the dyke rock and the coarse granite is greater than that of the fine granite, and so the features caused by the former are more pronounced. The coarse granite has been penetrated by veins of the finer granite and the lines of weakness along the junctions give rise to hollows by weathering. The sculpturing of the hill and valley system of Arran has all been accomplished within comparatively recent geological time, none of it being in existence before the Tertiary Period. Nearly all the prominent hills in the southern portion of the island are also formed of intrusive igneous rock of Tertiary Age, including Holy Island. Ard Bheinn, between Machrie Water and Ballymichael Glen, upon which the Tertiary volcanic neck already referred to is situated, is one of the most rugged of the Arran hills. This is due to the varieties of igneous rock that have invaded the neck.

Volcanic rocks of Lower Carboniferous Age occupy considerable areas in the Little Cumbrae and south Bute. In the latter locality they form a series of ridges which dip at high angles towards the south-west. In the Little Cumbrae the volcanic rocks are arranged in a shallow trough or syncline, forming a series of gently sloping terraces. Intrusive rocks of Carboniferous Age are numerous both in the Great Cumbrae and Bute, and give rise to prominent features such as the hill of Suidhe Chatain in the south of Bute.

Many of the smaller details of the Arran scenery are full of much interest. Reference was made on p. 28 to the recent views that have been advanced to explain the origin of corries, and the granite mountains of Arran contain many fine and instructive examples of these. The weathering of the granite along the joint planes has produced the so-called cyclopean walls of Goat Fell, in which the huge slabs of decomposed granite appear as if they had been piled up into a lofty mural precipice. The dykes, which are numerous along the shores of Arran, Bute, and the Cumbrae also present striking features, such as the "Lion Rock" and the "De'il's Dyke" near Millport. The action of ice in smoothing and polishing the asperities of the ground is well seen on Barone Hill, Bute. Moraines are abundant, especially in the Arran

**The Lion Rock, Millport**

glens. Travelled boulders are numerous, notably along the shore between Clachland Point and Glen Sannox. Landslips are numerous in Arran. Special mention may be made of the landslip of Upper Old Red Sandstone Conglomerate, known as the Fallen Rocks, on the shore four miles north of Corrie. It is recorded that the concussion shook the earth and the fall was heard both in Argyllshire and in Bute.

## 8. Natural History[1].

In the islands the usual sea-side plants occur, but in addition to these there is a group of plants having a peculiar Atlantic type. Amongst these are the following: the whorled carraway (*Carum verticillatum*), the Isle of Man cabbage (*Brassica monensis*), pennywort (*Cotyledon umbilicus*), English stonecrop (*Sedum anglicum*), the pale butterwort (*Pinguicula lusitanica*), Savi's mudrush (*Scirpus savii*), the sea spleenwort (*Asplenium marinum*), and the filmy fern (*Hymenophyllum tunbridgense*). Bute has also most of the common plants of the field, the roadside and the marsh, that are found in Britain. Among those less common in Scotland may be mentioned the marsh-mallow, the carline thistle, and the narrow-leaved white helleborine. Though the mountains of Arran reach to a considerable altitude, yet, when compared with the Breadalbane mountains, they are comparatively barren in Alpine plants. To this group belong such plants as the meadow rue (*Thalictrum alpinum*), the mountain lady's-mantle (*Alchemilla alpina*), the small enchanter's nightshade (*Circaea alpina*), the rose-root (*Sedum rhodiola*), four Alpine saxifrages—the *Saxifraga stellaris*, the yellow *S. azoides*, the purple *S. oppositifolia*, and the lady's cushion (*S. hypnoides*). Several rare rushes, sedges, and clubmosses also grow on the higher mountains. Seaweeds are abundant along the shores of the Firth, and many beautiful species may be collected at Lamlash and Whiting Bay.

[1] See p. 28.

Most of the mammals that occur on the mainland of Argyll also occur in Bute. Of the land birds the most conspicuous and important, especially in Arran, are the red and the black grouse, both of which are very abundant. Ptarmigan are also occasionally found near the summits of the granite mountains. Pheasants, introduced into Arran early the last century, are now numerous. The golden eagle is still occasionally seen on Arran. Falcons, hawks of various species, and owls are also numerous. The ring dove, the rock dove, the cuckoo, the swallow, the martin, the sand martin, the swift, the missel thrush, the common thrush, the redwing, the fieldfare, the blackbird, the whinchat, the redbreast, the hedge and the house sparrow, the yellow-hammer, the common bunting, the linnet, the chaffinch, the white, the grey and the yellow wagtail, and the wren are all common. The rarer land-birds include the kestrel, the goatsucker, the ring ousel, the water ousel, the wheat-ear, the golden-crested wren and the goldfinch. Among the water birds that frequent the coast are the oyster-catcher, the cormorant, the shag, the solan goose, the wild duck, the teal, the wild goose, the razorbill, the puffin, the northern diver, the common gull, the silver gull, the guillemot and the tern. Some of the rarer insects that have been found in Arran are the clouded yellow butterfly, the Scotch argus, and several species of the handsome fritillaries. What has been said about the marine zoology of Argyll may also be applied to Bute.

## 9. Along the Coast.

Starting at Lamlash, Arran, we find ourselves on the shores of a beautiful bay. Protected by the cone-shaped Holy Island, it affords a fine harbourage and usually contains numerous yachts and trading craft. King's Cross Point, opposite the lighthouse on Holy Island, is said to have been the point from which Bruce embarked for his expedition into Carrick. Round the point and a little further south is Whiting Bay with its pretty village. Crossing the mouth of Glen Easdale, we reach Dippin Point, near which is a basaltic causeway resembling the Giants' Causeway but on a much smaller scale. Further on is Kildonan Castle, and next the harbour of Drimlabarra. Pladda Island with its two lighthouses now comes in sight. The Struey Rocks and Bennan Head form the most striking features of this coast, the latter rising 457 feet. The Black Cave in front of the headland has been excavated in a vein of rotten trap between two basaltic walls. Proceeding westwards we pass Torrylin Waterfoot, near which stands the old church of Kilmory. At various places in the vicinity of Lagg beds of arctic shells may be seen. After crossing Sliddery Water we reach Brown Head and further on Drumadoon Bay and Drumadoon Point, consisting of quartz porphyry of perfect columnar structure. On the top of the Point are the ruins of an ancient fort. North-west from Drumadoon Point high cliffs of yellow sandstone extend along the shore, with numerous caves excavated by the sea when the land stood at a lower level. In one of these, the King's Cave, Bruce is supposed to have sheltered for a time when he first landed in Arran. Three miles beyond Machrie Water, we reach the opening into Glen Iorsa and obtain a fine view of the boulder-strewn mountains with the hamlet of Auchencar on the slope of

Drumadoon Point

Beinn Lochain. The road now skirts the shore at the base of the western slopes of Beinn Bharrain, passing through several hamlets and rounding Catacol Bay at the foot of Catacol Glen. A little further on Loch Ranza is reached, stretching inland for about a mile and about half a mile broad. The village consists of a few houses placed round the head and east side of the loch. In the foreground is

Loch Ranza Castle on a grass-covered peninsula, which runs far out into the bay. Rounding the extreme north end of the island we reach the Cock of Arran, a large isolated mass of sandstone resting on the beach, a noted landmark among sailors. Passing the "Fallen Rocks," we reach the openings of the Sannox Glens, with a bay and a hamlet at the foot of the more southerly. About a mile further on is the village of Corrie, with a small harbour. Five and a half miles south, we arrive at Brodick situated on Brodick Bay. The bay is 2½ miles wide between Merkland and Corriegills Points. North of the pier is a smooth sandy beach, and behind it a stretch of level ground, upon which the village stands. Behind the plain rise the granitic peaks, with Goat Fell towering above the rest and its skirts descending to the sea. On the north side of the bay stands Brodick Castle. Between Corrie and Brodick are low cliffs consisting of sandstone much traversed by dykes. Caves occur at many points, none of great size. From Brodick to Clachlands Point considerable cliffs approach close to the sea, composed mostly of red sandstone. At Corriegills a huge boulder, estimated to weigh 210 tons, rests upon the sandstone. It must have been carried to its present position from the granite mountains by ice. Further south is the celebrated vein of pitchstone conspicuous in the front of the sandstone cliff. After rounding Clachlands Point, we enter Lamlash Bay.

Crossing now to Bute, we take Rothesay as a convenient starting point. Rothesay Bay lies between Ardbeg Point on the west and Bogany Point on the east, which are about $1\frac{5}{8}$ miles apart. The bay is surrounded by gentle slopes beautifully diversified by woods and belts of trees. About three miles north-west of Rothesay is Kames Bay, which sweeps round in a half-moon form and has a good bathing beach. Round the north end of the island as far as Etterick Bay the

The Corriegills Boulder, Arran

shores are almost everywhere low and decided cliffs are absent. There is a flat margin along which runs the road. Scalpsie Bay is flanked on its southern side by the 25 feet raised beach. A tract consisting of a sea marsh and a sandy flat separates the extreme southern end of the island, terminating in Garroch Head, from the part that has just been described. This part consists of a series of ridges that dip towards the south-west and present their bold escarpments to the north-east. The west shore of Bute from Garroch Head to Etterick Bay is more picturesque and varied than the east side for the same distance. The coast line of the Cumbraes requires no detailed description. In the Great Cumbrae near

Raised Beach near Corriegills

Millport some of the dykes have resisted the denuding action of the sea, which has removed the softer sandstone on each side of them. The 25 feet raised beach also makes a prominent feature all round the island. The coast line of the Little Cumbrae is more picturesque than that of the Great Cumbrae, this being due to the weathering of lava beds into a series of terraces that rise in succession from the water to a height of about 409 feet. The Lighthouse stands on the lowest of these terraces.

At various places in Arran and Bute traces of raised beaches are found, from the 100 feet beach down to the latest at about 15 feet above the level of the sea. The 25 feet beach forms a conspicuous and picturesque feature round the islands, and most of the coast towns have been built upon it.

## 10. Climate[1].

The predominant winds in Bute as observed at Rothesay are from the east, south, south-west and west. Owing to the direction of the prevailing winds in Bute, rain is frequent, often copious and of long duration. The reason for this is not far to seek, for these comparatively warm winds blowing over an immense surface of water bring with them a large amount of aqueous vapour. On reaching such high land as we have in Arran, the air is forced to move upwards as well as forwards. The air is thus cooled by expansion and its power of carrying aqueous vapour is lessened. The vapour is then precipitated as rain. The average rainfall at Rothesay taken over a period of 100 years is 48·36 inches, and if this be compared with that of Islay (p. 48), it will be seen to be somewhat similar. At Dunoon the average rainfall is a little over 60 inches, so that we have in Bute much

[1] See p. 45

the same conditions as have been described for the seaboard of Argyll. The temperature at Rothesay in winter is 13° above the average in Scotland and in summer 5° cooler. Owing to the mildness of the winter climate, many of the trees that have been introduced from warm countries thrive well.

## 11. People—Race, Language, Population.

The earliest inhabitants of Bute belonged to the Neolithic or New Stone Age, who are considered to be the same as the men of the Danish kitchen middens. A midden, probably of this age, was found at Glecknabae in Bute on the old sea beach about 150-yards from the present shore; at Machrie Moor, Arran, the remains of stone circles are still to be seen.

The people who built the chambered cairns are the oldest inhabitants of Bute of whom we have any definite record. These people belonged to the later Neolithic period, and are supposed to have come from the south up the western coasts of Europe. Physically they were a short-statured race, with long, narrow heads and high, narrow faces without any projection of the jaws. They were of a dark complexion and not uncomely.

In the Bronze Age a new people came to Bute with different customs and culture. Each chambered cairn of the Stone Age was the burying place of several persons, but in the Bronze Age each short cist was a separate grave. These cists have been unearthed at Scalpsie Bay and West Lodge, Mountstuart, in Bute, and elsewhere in the county. The men of the short cist were taller than the men of the chambered cairns. Their skulls were short and broad, their faces broad and low.

In the early centuries of our era the inhabitants of

Stone circle on Machrie Moor, Arran

Bute and Arran were most likely Picts, but after 500 these islands were overrun by Dalriad Scots from Argyll. The Cumbraes, lying close to the coast of Strathclyde or Cumbria, were peopled by men of British race from Cumbria. The Norse designation *Kumreyar* indicates that when the Norsemen came they found the islands still in the possession of the Cymry. History has much

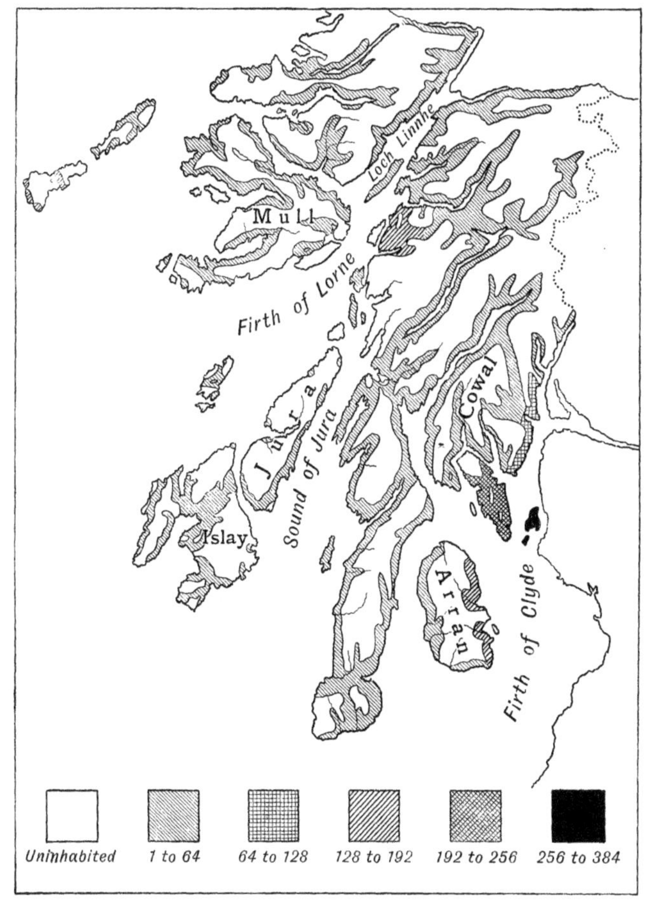


Map showing distribution of population to the square mile in Argyllshire and Buteshire

to tell us of Scandinavian raids on the islands of the Clyde and Scandinavian settlements there. Antiquarian finds prove the same. Norse elements appear in such names as *Rothesay*—older spelling *Rothersay*, "Rother's island" —*Brodick*—in Barbour's *Brus* it is *Bradwick*, "broad bay"—*Ascog*, *Scalpsie*, *Pladda*, *Ranza*, *Sannox*, *Goat Fell.* Norse and Celtic, then, formed the basis of the population in the Middle Ages. The men of Bute used to be styled Brandanes from the Irish sea-roving Saint Brandan (in the Old Irish, Brennain). See Matthew Arnold's poem. The saint's name appears in Kilbrannan Sound.

The distribution of the population in Buteshire does not seem to have much connection with its geological structure other than that the villages are usually confined to the coasts and the lower parts of some of the glens. In Arran the only exception to this is the district of Shiskine on the west side, where there are a few scattered farms at some distance from the sea. In Bute and Cumbrae farms are found scattered through the interior. The census of Bute taken in 1801 showed the population to be 11,791; there was a gradual increase till in 1851 it reached 16,608. In the following ten years the numbers slightly decreased, and in 1861 the population was 16,331. There was again a gradual rise and in 1901 it reached its maximum, 18,787. The following ten years show the second fall since the census was first taken, the numbers for 1911 being 18,186, or 601 less than in the previous census. Of the population, 2079 were returned as able to speak both Gaelic and English, while two were able to speak Gaelic but not English.

## 12. Agriculture.

About one-third of the whole county is unprofitable and little more than one-sixth is under cultivation. The agricultural statistics show that during the last half century great advances have been made by means of reclamation, draining, and the adoption of the best means of husbandry. The alluvial flats along the courses of streams and the raised beaches at their mouths afford the best soil. The terrace or raised beach round the islands is also carefully cultivated. Oats, turnips and potatoes are the staple produce of the shire. Horses, sheep and cattle are also reared. It used to be the general custom in the island of Arran to take the cattle to mountain pastures in the summer time, and remains of the summer shieling or *airidh* are common in nearly all the high glens. At Balliekine the old runrig system—under which alternate ridges in a field belong to different persons—common a hundred years ago, is still practised.

Arran has much natural wood, principally birch, alder, hazel, rowan and willow, with some scrubby oak; on the west coast, for example, from Dougrie and Loch Ranza, and on the east between Sannox and Brodick. Natural wood also grows in the lower parts of a number of the glens, as in Glen Cloy and Glen Rosa, to an altitude of about 800 feet. Along some of the smaller streams, especially with deep ravines, trees flourish up to 1000 feet; and in some instances the rowan tree has been observed at an altitude of 1500 feet. Artificial fir plantations are numerous in Bute and there are also a few in the Great Cumbrae. The largest in Arran is that around Brodick Bay, where the trees flourish up to nearly 700 feet.

Arran used to support a large crofter population. In the Millstone Point district in the north-east of the island 14 families resided at Cock, Cuithe, Laggan and Laggantuin; where there are now but a farmer and a shepherd. At one time a large population inhabited north Glen Sannox, where there is now but a solitary shepherd's house. In the Corrie district deserted farmsteads indicate former cultivation, where there is now nothing but pasture land.

## 13. Industries.

At one time Rothesay was the centre of several industries. About the year 1750 the manufacture of linen was introduced into the town but it did not flourish. What is claimed to be the first cotton manufactory in Scotland was started in 1779 by an English company, and afterwards passed into the hands of the celebrated David Dale, cotton spinner and philanthropist. For three-quarters of a century the industry flourished, employing four mills and 800 hands with 1000 looms, but it declined and ultimately died out. Tanning also flourished in Rothesay, but is now extinct. At one time there were two boat-building yards in Rothesay, but this industry too has disappeared.

Rothesay

In former days herring-fishing at Rothesay was encouraged by Government, which had, however, for its real object the obtaining of recruits for the naval reserve. In 1855 this industry employed 557 boats, 1654 fishermen, and indirectly 1102 other persons; 5074 barrels of cured herrings besides those uncured were sold. The industry, however, gradually drifted to the larger fishing centres which had greater facilities for curing the fish. In Arran also fishing is almost extinct, and is chiefly confined to catching herrings on the west coast of Loch Ranza and Pirnmill. The same may be said about Millport in the Great Cumbrae, where only a very few of the inhabitants carry on fishing. Rothesay is a port of some importance. The exports are principally the agricultural produce of the island while the imports consist of miscellaneous small goods. These are carried by steamers plying from Greenock, Gourock, Wemyss Bay, Helensburgh and Ardrossan to Rothesay, Millport, Brodick and Lamlash.

## 14. Mines and Minerals.

Coal was formerly mined in Arran near Cock Farm at the north end of the island. The coal was of the blind or glance variety. Two or three seams are said to have been worked, the most important being three or four feet thick. The workings are very old, as no coal has been mined in the locality for over a hundred years. Coal has also been mined at Ambrisbeg near

Scalpsie Bay, in Bute, some 200 or 300 tons of coal having been raised many years ago. A similar seam was worked at Ascog. The coal, a kind of anthracite, was locally known as blind coal. Blain's *History of Bute* states that the Ambrisbeg seam varied from a foot to fifteen or twenty inches in thickness. As the coal is thin and the area of the field small, it cannot be considered to be of much economic value.

Peat is abundant in Arran, especially on those tracts which rise to between 700 and 1700 feet, but it is occasionally found at still lower levels. Up till recent times peat was the principal fuel in the Highlands and Islands, but the increased facilities for the transit of coal from the south have led to its gradual disuse. Old peat roads are to be met with almost everywhere on the Arran hills. A thin band of clay ironstone crops out on the shore at Corrie and a similar one further north at the Cock. The ironstone never seems to have been mined, but simply to have been gathered during the working of the freestone quarries at Corrie. At one time some hundreds of tons of it were exported. Evidence of the existence of old bloomeries is to be found in various parts of Arran. One of these on the farm of Glenkiln near Lamlash shows that bog iron ore was used for the production of the metal. The fuel used in every case seems to have been charcoal. The slag is of the dense black type characteristic of the early period in smelting the ore when as yet the process was but imperfectly applied. Barytes has been worked in Glen Sannox about a mile from the sea on the north side of the burn. The mineral occurs in a massive form, generally of a pinkish or yellowish tinge, but sometimes almost pure white. The white Carboniferous Sandstone of Corrie was worked for building purposes about a century ago. It is said to have been used in the construction of the Crinan Canal and to have been shipped to the Isle of Man. The

Corrie limestone, Arran

red sandstone of Triassic Age, quarried at Corrie and Brodick, is largely exported from Corrie to different parts of the Clyde and the West Highlands. It is said to make a very durable building stone. White freestone of Upper Old Red Sandstone Age is quarried at Millport.

Slate was formerly quarried in Bute at Kames, on the island of Inchmarnock, and south of Cock Farm in

Arran. It was, however, of a very rough nature and could not compete with the slate of Easdale and Ballachulish. Limestone has been worked at Corrie in Arran and at Kilchattan in the Island of Bute. At Kilchattan there is a fine laminated clay belonging to the glacial series, which has been extensively worked for tile-making.

## 15. History of the County.

From the fifth century to the eighth, consequent upon the coming of the Dalriad Scots to Kintyre and in the contests among the Scots of Argyll, the Picts of the North, and the Britons or Cymry of Strathclyde, much fighting took place in and around the Firth of Clyde. In these struggles Bute, Arran and the Cumbraes, from their strategical positions, could not fail to share; but details are extremely uncertain. Then when the Norsemen swooped down, whether in a summer raid or to make a permanent settlement, the islands of the Clyde were the scenes of many a stern conflict. Olave and Ivor, for example, two kings of the Northmen from Dublin sailed past these islands on their way to besiege Dumbarton Rock, which they took after a four months' siege in 870. Magnus Barefoot was here in 1098 during his campaign to scourge the disobedient chiefs of the isles, and is said to have then begun the building of Rothesay Castle. Here also took place the last struggle between the Norwegians and the Scots in 1263. Haco of Norway had come with 160 ships of war to maintain his claim to his old possessions, and had occupied Bute and Arran. He attacked and captured Rothesay Castle. The Scots offered to let Haco keep the outer islands if they might hold the mainland with Arran, Bute and the Cumbraes. Haco refused. October gales shattered his fleet; and,

The King's Cave and Drumadoon Point, Arran

defeated in battle at Largs, he sailed away north to die in the Orkneys.

When Bruce was fighting for the Crown, Brodick Castle was held by the English under Sir John Hastings, the third "competitor" for the throne along with the elder Bruce and John Balliol. Sir James Douglas snatched the castle from his hands. Soon after this, Bruce in 1307 arrived on the west of Arran from his retreat in Rathlin. Tradition makes him lurk in a cave north of Drumadoon Point. Then traversing the island, he waited at King's Cross Point, at the south end of Lamlash Bay, for the fire on the Ayrshire coast that was to be the signal that the time was propitious for an attack on Turnberry Castle. See Scott's *Lord of the Isles.*

**Rothesay Castle**

In the troublous days after Robert the Bruce's death, when Edward Balliol sought to win the throne, Rothesay Castle was at one time in the hands of Balliol, at another in the hands of David Bruce's supporters.

Bute, like Renfrew, became the patrimony of the Stewarts; and it was a favourite residence of Robert II and Robert III. In 1398, when the dignity of duke was first created in Scotland, Robert III's son, David, Earl of

Carrick, was made Duke of Rothesay—a title which the monarch's eldest son has ever since held. One of the Scottish heralds is styled Rothesay, and one of the pursuivants Bute. Rothesay Castle was very probably the place where Robert III died, though one authority says Dundonald in Ayrshire. After the battle of Dunbar, some of Cromwell's soldiers came to take Rothesay Castle, which suffered considerable damage. But it was in 1685 that it was reduced to a ruin, in connection with the Earl of Argyll's futile insurrection in favour of Monmouth. The Earl's second son, Charles Campbell, was sent to occupy Bute as a suitable base of operations, where also food and men could be got. His Highlanders plundered houses, robbed the inhabitants of cattle and money; they even emptied the poor's box in the Parish Church of Rothesay. By their leader's orders they burned down the part of the Castle that was still habitable.

## 16. Antiquities.

Arran and Bute are rich in antiquities belonging to the Stone and the Bronze periods.

One of the best preserved of the chambered cairns of the Stone Age stands at the head of Kilmory Water in Arran. It consists of a rectangular heap of stones which appears to have been at one time marked off by a continuous row of flagstones set on end. At the western end the cairn has a semicircular or bay-like indentation and in the centre of this stand the portal stones which give entrance to a chamber or vault roofed in by a series of great flags. The chamber, which is 18 feet in length, is a narrow trench separated into compartments by cross stones. The walls consist of blocks of stone set on edge and placed parallel to one another. Upon the uneven edges of this basal part smaller stones had been piled till the required height was reached and a flat edge was formed upon which to rest the flags of the roof. Professor Bryce, who has made a detailed examination of these cairns, says that seventeen examples of them exist in Arran, while others are to be found in Bute to the west of Loch Fad. The implements found in these chambers are all of stone and include flint knives, arrow heads, stone axes, and hammer heads. It is interesting to note that flakes and blocks as well as arrow heads of the Corriegills pitchstone have been found, showing that in Buteshire pitchstone had been used as a substitute for flint. Round-bottomed bowl-like vessels have also been found in these chambers.

Urn from Cumbrae

A tumulus which was opened at Scalpsie Bay in the island of Bute, belongs to the late Bronze Age. The centre was found to consist of a core of stones, and upon their removal a large flagstone was revealed, which covered a stone cist. The cist was formed of four neatly dressed flags enclosing a space 2 feet 10 inches by 1 foot 6 inches, and contained a beautiful urn, a flint knife, some beads and a bronze pin. The bronze pin indicates that the interment belongs to the Bronze Age, while the flint knife shows that stone implements had not been entirely disused. The bones were burnt, proving that the body had been cremated.

Cup-and-ring markings at Brodick

In a cist at Blackwaterfoot in Arran a beautiful bronze dagger-blade was found with gold fillets, probably portions of the mounting of the dagger. These cist burials are found in cairns or mounds which are sometimes surrounded by standing stones, as at Tormore and other places in Arran, and at Etterick Bay and Kilchattan in

Bute. In other cases no monument has been raised to mark their position. Cup-and-ring-marked sculpturings, sometimes on separate boulders but also on the native rock, have been found in Arran at Stronach Ridge. Archaeologists have come to no certain conclusions about the age, the origin, and the significance of these markings.

A Viking grave-mound was recently found on the south of Lamlash Bay, at King's Cross Point. Another Norse monument is a cross with runic inscription in the old churchyard at Inchmarnock.

## 17. Architecture—Ecclesiastical, Castellated, Domestic.

On the Holy Isle, Arran, is the hermitage of an anchorite St Molios, from which the island has received its name. It is a natural cave in the old sea cliff and recent investigations have revealed a stair, pavemented floor, and fireplace, so that it seems quite likely that we have here the abode of one of the early missionaries. About a mile from the village of Lamlash is the quaint old pre-Reformation chapel of St Bride—now in a ruinous state—which probably belongs to the fourteenth century. There were other chapels at Kilmichael, Sannox, Loch Ranza and Kilmory, but little or nothing now remains of these, with the exception in some cases of traces of the foundation stones. Within the court of Rothesay Castle stand the walls of the ancient chapel of St Michael. It may have been erected in the fourteenth century, when Robert II and Robert III frequently visited the Castle.

Loch Ranza Castle is situated at the north end of Arran. It consists mainly of two roofless square towers united, but evidently belonging to different periods. The thick and massive walls and the position of the Castle show that it must have been a place of great strength and security. Little is known of its history, but it was

**Loch Ranza Castle and Harbour**

a royal hunting-seat in 1380. Kildonan Castle, at the south-eastern extremity of Arran, stands on a precipitous sea-cliff. The castle is a square keep, without ornament, four storeys in height with several vaulted apartments. It was probably built during the wars with the English Edwards and formed one of a line of watch-towers from

Ailsa Craig to Dumbarton Rock. It was a royal residence up to the year 1405, when it passed into the hands of Stewart of Ardgowan, in whose family it remained along with the lands attached for 150 years. Rothesay Castle consists of a circular building 138 feet in diameter with walls 8 to 10 feet thick and 26 feet high. Of the four round towers which flanked the main building only the one to the north-west remains in good preservation, the three others being considerably demolished. They have each a door opening from the courtyard on to the ground level. The building which runs to the north of the ancient entrance tower and which forms the present entrance, is of a later date. The castle became a complete ruin in 1685; but was repaired, 1871–77, by the Marquis of Bute.

Brodick Castle, formerly one of the seats of the Dukes of Hamilton, stands on the north side of Brodick Bay. It is in the Scottish baronial style with a lofty tower at one of the corners, which is flanked with small turrets and capped with crow-stepped gables. The castle is beautifully situated and forms a fine feature in the landscape of the bay. Mountstuart House, the magnificent mansion of the Marquis of Bute, stands about five miles south of Rothesay. It was begun in 1879 and is a fine Gothic pile with high pitched roofs, angle turrets and corbelled oriel windows. The central hall, 60 feet square, is surrounded by a Gothic arcade of marble, from which the dining room and principal drawing room lead off to right and left. The house is approached by a fine avenue, and stands in extensive and carefully laid out grounds.

Kames Castle, also belonging to the Marquis of Bute, is at the base of Kames Hill close to Kames Bay. The main portion of the building consists of a large mediaeval tower, which from its style of architecture is supposed to have been built in the fourteenth century. It was the birthplace of John Sterling, critic and essayist.

## 18. Roll of Honour.

Bute has a comparatively small roll of honour. John Stuart—"Jack Boot" of the London mob—Earl of Bute, born in 1713, was elected a representative peer of Scotland in 1737, but preferred botany and agriculture in Bute to politics. Some ten years later he became attached to the court of Frederick, Prince of Wales, on whose death he was appointed Groom of the Stole to Prince George, afterwards George III. Bute influenced the young Prince very much and in 1761 was one of the principal Secretaries of State and next year Prime Minister. His incapacity and inexperience, his aim of making the king absolute, and his Scottish origin made his government a failure and himself detested. His last years he spent in retirement, busied with botany and other studies. His great-great-grandson, John Patrick Crichton Stuart, Marquis of Bute (1847–1900), was deeply interested in religious and philanthropic movements. He was a benefactor of Glasgow and St Andrews Universities. His writings include an English translation of the *Breviary*, and other ecclesiastical works.

John Stuart, Earl of Bute

Dr Matthew Stewart, born at Rothesay, studied at Glasgow and Edinburgh and became minister of Rosneath. In 1746 he was made Professor of Mathematics in Edinburgh, and wrote *General Theorems* and *Tracts, Physical and Mathematical.* His son was the famous philosopher, Dugald Stewart. Daniel Macmillan, born in Arran, at Upper Corrie, in 1813, became a bookseller in Irvine. Finally settling in London, he along with his brother founded the firm which developed into "Macmillans." John Sterling (1806–1844) was born in Kames Castle, where his father was then living. He studied at Glasgow University and at Cambridge—where he was one of the "apostles"—conducted the *Athenaeum* for a few months, and wrote for various magazines. In 1835 his friendship with Carlyle began. Sterling is best remembered by Carlyle's biography of him.

## 19. THE CHIEF TOWNS AND VILLAGES OF BUTESHIRE.

(The figures in brackets after each name give the population in 1911, and those at the end of each section are references to pages in the text.)

**Ascog** (118), overlooks the Bay of Ascog about $1\frac{1}{4}$ miles south of Bogany Point at the entrance to Rothesay Bay. It

Brodick Bay, from Dun Dubh

consists of a chain of cottages and mansions extending for two miles along the shore. (pp. 136, 141.)

**Brodick** (716), a favourite watering place and resort of families from Glasgow, on Brodick Bay, amid most varied and beautiful scenery. It is a capital bathing and boating place. (pp. 105, 109, 115, 129, 151.)

**Lamlash** (634), a retired village and summer resort at the head of Lamlash Bay. The parish church of Kilbride is a small but beautiful edifice erected at the expense of the Duke of Hamilton in 1884. (pp. 127, 140, 141, 145, 149.)

**Millport** (1614), on a pleasantly sheltered bay at the south end of the Great Cumbrae, is built in a crescent following the curve of the bay. Notable buildings are the parish Church with its low square tower; the Garrison, a marine pavilion; the Episcopal College; and the Collegiate Church or Cathedral, founded in 1849 by the Earl of Glasgow, in the thirteenth-century Gothic style. At Keppel Pier is the Marine Biological Station and Museum instituted for the purpose of investigating the marine zoology of the Firth of Clyde. (pp. 106, 109, 124, 131, 140, 142.)

**Port Bannatyne** or **Kamesburgh** (990), a village on the southern shore of Kames Bay, 2½ miles north from Rothesay. It was formerly a fishing station but this industry has greatly declined. The village is overlooked by Kames Hill, the highest in Bute. Half-a-mile to the north is the ancient Castle of Kames. (pp. 142, 152, 154.)

**Rothesay** (9299), most picturesquely situated on Rothesay Bay, is the chief town of the county and the most popular resort on the Clyde coast. Its terraced slopes, curving shores and mild climate have earned for it the title of "the Madeira of Scotland." The steamboat quay is a substantial stone structure with two harbour basins, and is the busiest on the Clyde. The town is a most excellent centre for visiting all places of interest on the Firth of Clyde. Rothesay was made a royal burgh in 1400. (pp. 105, 106, 109, 117, 129, 132, 138, 140, 143, 145, 149, 151, 154.)

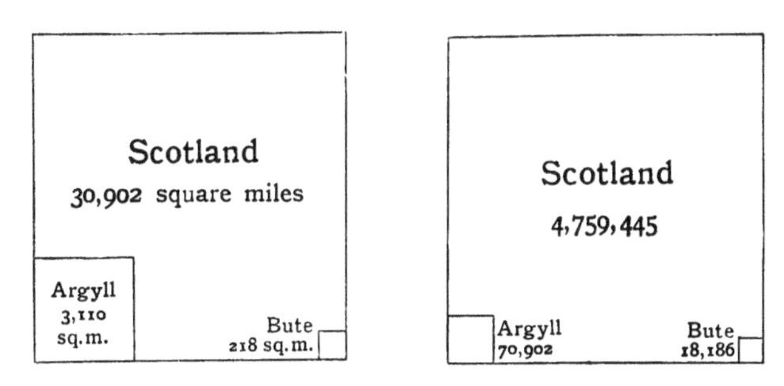


Fig. 1. Comparative areas of Argyllshire and Buteshire and all Scotland

Fig. 2. Comparison in population of Argyllshire and Buteshire and all Scotland

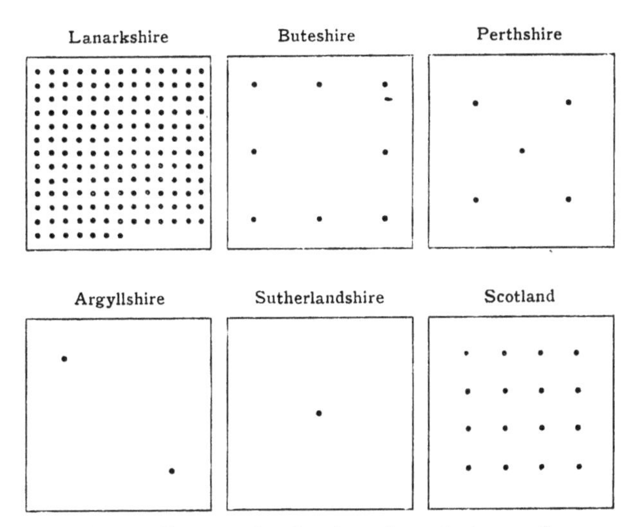


Fig. 3. Comparative density of population to the square mile in 1911

*(Each dot represents ten persons)*

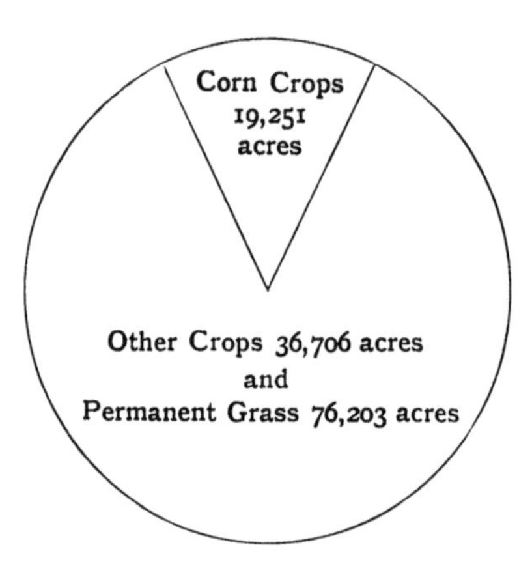


Fig. 4. Proportionate area under Corn Crops compared with that of other land in Argyllshire in 1911

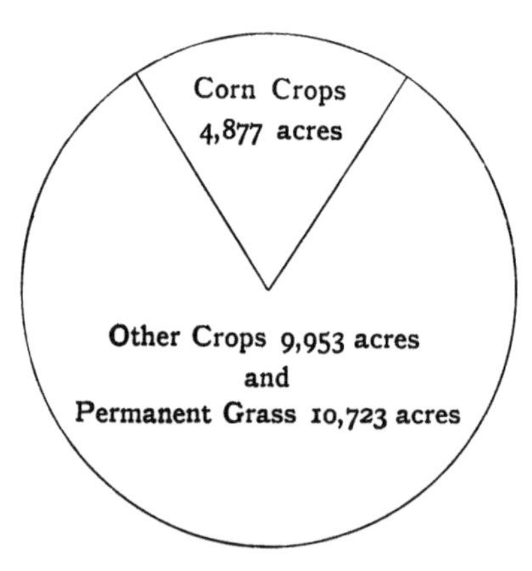


Fig. 5. Proportionate area under Corn Crops compared with that of other land in Buteshire in 1911

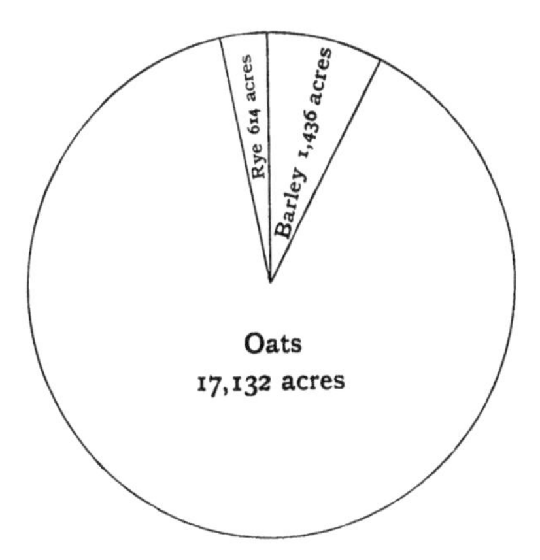


Fig. 6. Comparative areas under Cereals in Argyllshire in 1911

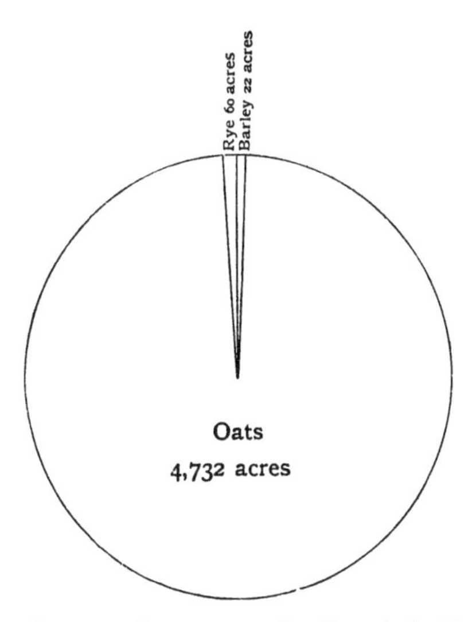


Fig. 7. Comparative areas under Cereals in Buteshire in 1911

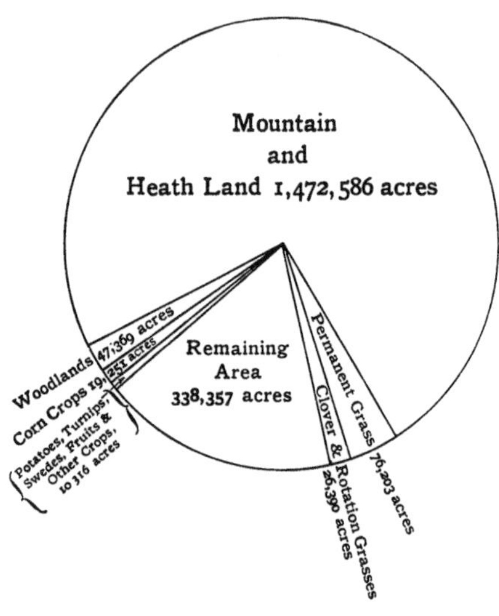


Fig. 8. Comparative areas of land in Argyllshire in 1911

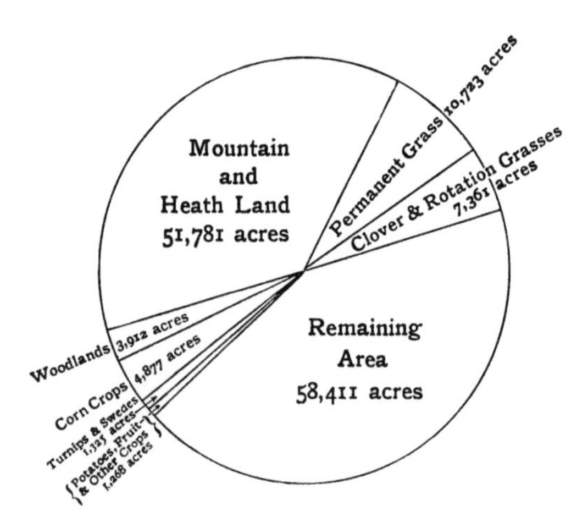


Fig. 9. Comparative areas of land in Buteshire in 1911

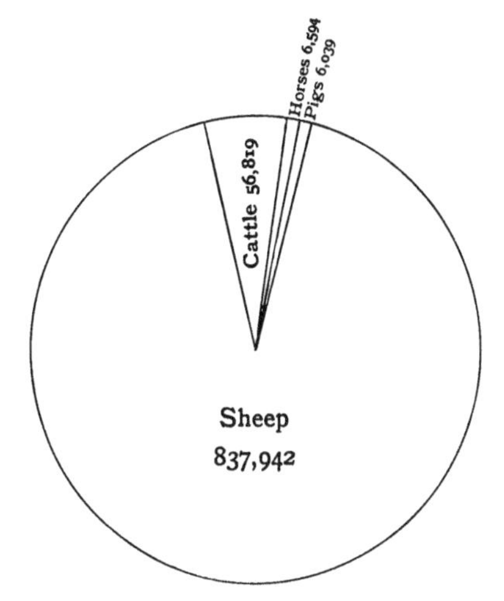


Fig. 10. Comparative numbers of different kinds of Live Stock in Argyllshire in 1911

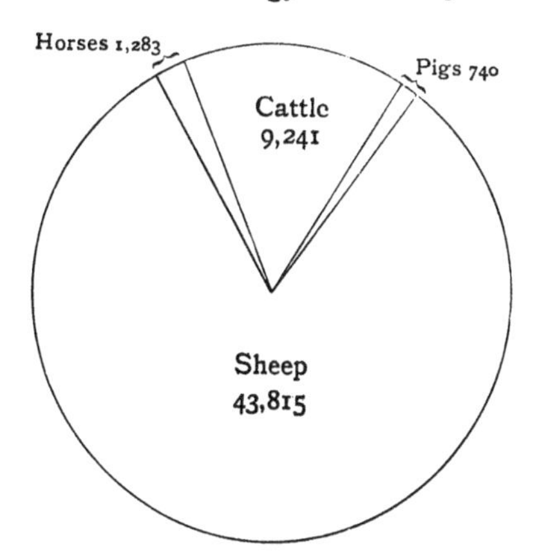


Fig. 11. Comparative numbers of different kinds of Live Stock in Buteshire in 1911

www.ingramcontent.com/pod-product-compliance
Ingram Content Group UK Ltd.
Pitfield, Milton Keynes, MK11 3LW, UK
UKHW042144280225
455719UK00001B/95